Lavery Library

EX LIBRIS
ST. JOHN FISHER COLLEGE

D1456473

THE INTERNATIONAL REGULATION
OF EXTINCTION

Also by Timothy M. Swanson

ECONOMICS FOR THE WILDS (*editor with Edward B. Barbier*)

ELEPHANTS, ECONOMICS, AND IVORY
  (*with Edward B. Barbier, Joanne Burgess and David Pearce*)

The International Regulation of Extinction

Timothy M. Swanson

Lecturer in Economics
University of Cambridge

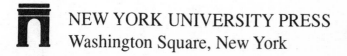

NEW YORK UNIVERSITY PRESS
Washington Square, New York

© Timothy M. Swanson 1994

All rights reserved

First published in the U.S.A. in 1994 by
NEW YORK UNIVERSITY PRESS
Washington Square
New York, N.Y. 10003

Library of Congress Cataloging-in-Publication Data
Swanson, Timothy M.
    The international regulation of extinction / Timothy M. Swanson.
    p.   cm.
    Includes bibliographical references and index.
    ISBN 0–8147–7992–1
    1. Biological diversity conservation—Government policy.
2. Biological diversity conservation—International cooperation.
3. Economic development—Environmental aspects.   4. Extinction
(Biology)   I. Title.
QH75.S815   1994
333.95' 11—dc20   93–36643
                                                              CIP

Printed in Great Britain

Contents

List of Tables

List of Figures

Preface

This book covers a lot of ground in a number of different ways. It is an analysis of evolutionary history, a process entailing approximately four billion years of life on earth. It is also an analysis of human societal development, a process occurring over about ten thousand years and diffusing to the ends of the earth. As these are both global phenomena, they require international forms of regulation. The analysis within this book must therefore be both broad and interdisciplinary, incorporating legal, economic, and ecological facets within a wide-ranging framework of global resources in development over the very long run.

Despite the over-ambitious scope of the book's content, there is a simple structure to the volume, that may appeal to diverse groups of students and scholars. The first half of the book may be seen as a revised economic theory of extinction, incorporating the ideas of biodiversity decline and institutional underinvestment within the existing theory. Chapter 3 provides the general statement of this theory. Chapter 4 concentrates on the aspects of this theory concerned with diversity depletion, while Chapter 5 focuses on the institutional factors contributing to extinctions. Together these chapters provide a new economic theory of extinction and biodiversity decline.

The second half of the book (Chapters 6 to 9) provides a depiction of the international regulatory problem pertaining to the forces outlined within the first half of the book. Chapter 6 details why the currently existing (or non-existing) system of international regulation fails to control global diversity decline optimally. Chapters 7, 8 and 9 focus on specific components of an international system of regulation that might be introduced, for example under the auspices of the recently adopted Convention on Biological Diversity.

In total, the book spells out the problem and the policies with which international biodiversity policy must be concerned. Of course, it is perfectly permissible for the international community to cause the biodiversity convention to address any number of range of problems that it might desire, *but the point of this volume is that there is at least one clear and demonstrable deficiency in existing regulatory structures concerning extinction that must be addressed.* At the very minimum, this book argues, the Biodiversity Convention must be concerned with halting the conversion of the last great remaining areas of natural habitat, by encouraging the development of these areas for uses compatible with the natural diversity resident there.

This book is written in a format that will be readily accessible to scholars in the rapidly growing field of law-and-economics. However, given the range of disciplines interested in the subject matter, I have attempted to formulate an approach that is more accessible to those accustomed to other modes of presentation, although this has lent to the aforementioned weightiness of the tome. So, the presentation often discusses a point in literary form, and then develops the same point mathematically. This is meant to allow those who are acquainted with the former to skip over the latter, and vice versa. I apologise for the redundancy, but it seems unavoidable in a book of this nature and scope.

I have intended that the book be of use and of interest to the growing number of students and scholars interested in the interface between biological resources and social institutions. It may also be of interest to the various constituencies (governmental and non-governmental) involved in issues of conservation and development, if they have it translated for more general consumption by a friendly economist. In any event, I send it off in the hope that it makes some contribution to a constructive debate on the future shape and direction of international conservation policy.

TIMOTHY M. SWANSON
Cambridge

Acknowledgements

This book is the result of five years of continuous and almost exclusive work and thought on the subject of diversity decline. The complete roster of influential people and programmes would be immense. An abridged, and somewhat chronological, roster would read as follows. First, thanks are due to Professor David W. Pearce and our associates at the International Institute for Environment and Development, who introduced me to the Ivory Trade Review Group in 1988. It was this involvement that initially started the wheels rolling in this field. Secondly, I need to thank my colleagues at my current institutions: the University of Cambridge and the Centre for Social and Economic Research on the Global Environment (CSERGE) for the helpfulness and support needed to complete this research. Also, I should acknowledge the financial support of those institutions as well as that of the National Westminster Bank which made the accomplishment of this work possible. Third, I would like to thank those individuals who have commented extensively on previous drafts of this work, especially David W. Pearce of CSERGE and Bob Rowthorn of Cambridge. I would also like to thank those individuals who have involved me on the policymaking side of conservation policy, especially Jorgen Thomsen of TRAFFIC International and Steve Edwards of IUCN. Also, I must express my debt of gratitude to Professor Nick Stern, of the London School of Economics, who has put a lot of his very scarce time into this project. I have benefited immensely from his willingness to endow me with his time and with his insights. Finally, I must thank all those Macmillan staff who contributed to the completion of this book.

Although she would disclaim any responsibility for the initial undertaking of this project, I wish to dedicate the final product to my wife, Jean Marie, with my love forever. I hope that, despite all of the intervening years, the completion of the journey has rendered its undertaking worthwhile.

1 Introduction

1.1 INTRODUCTION: THE INTERNATIONAL REGULATION OF EXTINCTION

This is a book about the conservation of the diversity of life forms that exists on earth. That is, it is a book about evolved biological diversity, or 'biodiversity'. Biological diversity is important because it is the primary output from a four-and-a-half-billion year evolutionary process. The evolutionary process has endowed the earth with this range of diversity for important reasons, from the perspective of humans and for the preservation of life on earth. This book is about the conservation of this unique global endowment.

It is also a book about global development and North–South relations; and therefore deals with the 'uneven' character of global development. This is because the historical pattern of global development has resulted in a clear asymmetry in the world, between those states with material riches and those with species-richness. For example, a majority of the earth's biological variety now resides in the last, vast tropical forests existing in a few countries in the Southern hemisphere: the Latin American Amazon, and the Zairean and the Indonesian forests. Given this starting point, this book focuses primarily on the conservation of these last great wildernesses. And, given the rights of these states concerned to control their own destinies, and hence to control the resources within their borders, this book focuses ultimately on the resource allocation decisions of a handful of Southern states, their impact on the global environment, and the capacity of the global community to influence those (land use) decisions.

This book is also about the cumulative impact of global development; that is, it is about the expansion of the human niche, and the appropriation of many others. Over time, general global development has resulted in a systematic narrowing of the variety of life forms on earth, with the prospect of a much more severe narrowing in the next one hundred years; for example, many of the world's leading biological scientists are now predicting losses of a substantial proportion of all life forms on earth over the next century. In order to understand the nature of the problem of biodiversity depletion, it is necessary to examine how and why global development results in the systematic reallocation of basic resources away from a wide range of the world's species.

1

Primarily, however, this book is about the human institutions required for the management of biological diversity. It concerns the economic incentives that drive the human-made extinction process. It concerns the institutions (primarily domestic) that exist at present that do not adequately manage this process. Most importantly, it concerns the institutions (primarily international) that do not exist at present, but are necessary for the adequate management of biodiversity. These are the book's concerns.

The human management of the extinction process might sound like a typically anthropocentric perspective on a problem that concerns primarily non-human life forms. Or, at a minimum, it might appear to most to be a highly impracticable objective. However, the point of this book is that the human regulation of extinction is something that is already taking place -- only on a wasteful and inefficient basis. This is true, even from the human perspective. Therefore, this book does not make the case for exclusive human dominion over the other species of the planet; rather, it appeals for the reasonable exercise of that power which already exists. It appeals for this through the vehicle of human interest because it is human interests that drive extinctions at present, and these same interests may instead be channelled to the conservation (rather than the elimination) of diversity. (Also, human interests are preeminent in this book because the loss of diversity is likely to threaten human production systems, and hence human life forms, long before it constitutes a threat to life itself.)

Therefore, the human regulation of extinction is a system that already exists, but it is a poorly managed group of forces and incentives that appear set to generate a near-term mass extinction of species. The issue is whether this extinction process will be allowed to continue through human ignorance and avoidance of responsibility, or not. This book demonstrates the wasteful nature of the existing system of regulation, and indicates the direction for potential reforms for the conservation of biological diversity.

1.2   METHODOLOGY: THE BIOLOGY–ECONOMICS–INSTITUTIONS INTERFACE

There is little choice in a book of this nature but to create a broad inter-disciplinary framework for analysis. This is because the nature of the problems involved cannot be addressed in a context that omits consideration of any of the fundamental contributing forces to the management of the extinction process.

First, this subject requires the incorporation of biological analysis, especially the understanding of some evolutionary biology, because the biodiversity resource is essentially the output of the evolutionary process. That is, although each of the constituent resources of biodiversity is biological in nature (in that they grow and reproduce), the relevant characteristic of biodiversity is instead the range and variety of life forms, its 'diversity', rather than their biological character. It is the aggregation of the differences between biological organisms that matters in the management of biodiversity, and these differences are the output from the evolutionary process.

Secondly, it is necessary to incorporate economic analysis into the enquiry because, as indicated previously, it is the human drive for wealth and fitness that underlies much diversity depletion. In effect, the extinction process is now a distinctively social rather than a natural process. Human societies have usurped the evolutionary function of allocating base resources (i.e. those required for the sustenance of life) between competing life forms, and this is what now drives the extinction process. The fundamental nature of the biodiversity problem is a human choice problem, where human societies are systematically reshaping the biosphere across the face of the earth. Economic analysis lies at the core of the examination of this problem because it is the study of human choices, especially those choices concerning the human allocation of resources. This is the nature of the human-sourced extinction process, even though the issues involved here concern the allocation of resources by humans *to other species*.

Finally, it is most important to combine these two approaches with a third: the analysis of the institutions that provide the interface between human societies and biological resources. The evolutionary process endowed the earth with this resource. The human drive for wealth and fitness has placed it under threat. It is the institutions of human society that will determine the final outcome of this interaction between man and biosphere. In the event that the human drive for wealth and fitness cannot be tempered, then it will be the re-channelling of this force, in a manner made compatible with biological diversity, that will halt the process of diversity depletion. This is the role of human institutions – to manage constructively the forces of human society. The object of this book is to analyse what institutional reforms are required to manage the human drive to deplete biological diversity.

1.3 BIOLOGICAL DIVERSITY: A DEPLETABLE ENDOWMENT FROM THE EVOLUTIONARY PROCESS

Biological diversity, in the biologist's sense of the word, is the natural stock of genetic material within an ecosystem. This stock may be determined by the actual number of genes existing within the system. The number of genes range between organisms from about 1000 in bacteria to 10 000 in some fungi, and to around 100 000 for a typical mammal. The greatest number of genes actually belong to the flowering plants, whose genes often number in excess of 400 000 or more. Genes are important because they determine the particular characteristics of a given organism. They encode the information which determines the specific capabilities of that organism. The greater the variety in the gene pool, the greater is the variety of organisms which exist or which will exist in the near future.

The usual unit of analysis in studies of biodiversity is the number of existing species. This is because biological diversity may be conceived of as the net result of the processes of speciation and extinction. The number of currently described species number around 1.4 million; however, the number of species which have not yet been catalogued far exceeds that number. Best estimates place the total number of species somewhere between five million and 30 million. The vast majority of these species are insects, and other smaller organisms. Among the more well-studied categories, vertebrates and flowering plants, the numbers are much lower and more certain. For example, it is known that there are about 43 853 vertebrates currently in existence, of which only 4000 are mammals, 9000 are birds, 6300 are reptiles, and 4184 are amphibians. By way of contrast, it is known that there exist at least 50 000 different species of mollusc (Wilson, 1988).

This diversity of species has resulted from the process of 'radiation': the mutation of species and their consequent expansion into unoccupied niches. It is a process that has occurred over many hundreds of millions of years to produce the diversity this world currently contains (Raup, 1988). However, this process has never been a continuous or cumulative one. The fossil record indicates that there is about 4.5 billion years of earth's biological history. The record of the first 4 billion years, so-called 'deep time', indicates that very little adaptive radiation occurred over that very long period. Simple one-cell organisms (blue-green algae) were all that came to exist over the first two billion years of the earth's existence, and only very simple multi-cellular organisms are found to have evolved over the next two billion years.

At the beginning of the next half-billion years (ranging to the present and known as 'shallow time'), some unknown event triggered a revolution in adaptive radiation. The evolution of complex organisms, and their subsequent speciation, occurred all in a rush at the beginning of the Paleozoic era. Almost all of the major phyla presently existing appeared at that time, over a relatively short period of one hundred million years. The size and range of the biodiversity resource was determined by this revolution (Stanley, 1986).

Since that time the average rates of speciation and extinction have been approximately equal. Therefore, the amount of biological diversity existing at present, in terms of the number and variety of species, is probable very close to its historical maximum. Nature, at the beginning of the Paleozoic period, endowed the earth with approximately its present amount of diversity, and there is very little known about why this occurred.

Of course, extinction has itself been a natural process. Studies of the fossil record show that the natural longevity of any species lies in the range of 1 to 10 million years. The threat to biological diversity arises when the rate of extinction of species far exceeds the rate of speciation. In these eras of 'mass extinctions', there is the potential for a threat to the entire global biology. In the distant past, so-called 'deep time', there have been a number of occasions of such mass extinctions. There are at least five occasions indicated in the fossil record during which over 50 per cent of the then-existing animal species were rendered extinct (Raup, 1988).

Even averaging in these periods of mass extinctions, the natural rate of extinction over deep time appears to have been in the neighbourhood of 9 per cent per million years, or approximately 0.000009 per cent per year. That is, the current stock of biological diversity is the result of several billion years of mostly low-frequency mutation and extinction. Mass extinction has been an infrequent 'unnatural' occurrence brought on by exogenous shocks to the earth's biological system, such as collisions with meteors, volcanic activity, etc.

This brief account of speciation and extinction demonstrates the facet of biological diversity that is of the nature of a non-renewable resource. The *diversity* of biological resources is a one-time endowment from the evolutionary process. Therefore, although individual biological organisms may be treated as renewable resources, the aggregation of differences between these resources, i.e. the diversity that they represent, is best conceptualised as a non-renewable resource. Biodiversity is *the* natural resource which exists at the interface between the spheres of renewable and non-renewable resources.

It is also important to note why it is the case that human technology is incapable of resolving the problem of biodiversity depletion. Humans now have control of the rate of speciation as well as that of extinction, through the use of biotechnological methods. However, there is a very substantial difference between that variety which has developed through the process of evolution and coevolution over a period of six hundred million years, and that which can be created by experimentation in the laboratory.

The output from the evolutionary process, by the definition of evolution, has been selected for by reason of its capability to interact within the system, including its biological and chemical activity and its role in its ecosystem. That is, the evolution of a particular life form is an indicator of its capacity to act upon and contribute to the other organisms within its environment. Biodiversity is valuable precisely because it is the output from this four billion-year-old evolutionary and coevolutionary process, not merely for the sake of variety itself. Existing life forms are an encapsulated history of this process, and this constitutes an entirely unique body of information.

Another value of biodiversity lies in the fact that evolution has produced this particular range of variety. Evolution has a built-in capacity for adaptation, and the variety that exists is then an indicator of the requisite range of potential responses life has required for meeting changes in the physical environment. The range of existing life forms developed by the evolutionary process then constitute a uniquely formulated insurance policy against shocks to the life system itself, once again because existing life forms encapsulate a history of successful adaptation within a changing physical environment.

Therefore, in biological diversity we are dealing with one of the ancient non-renewable resources, such as the fossil fuels, rich soils and great aquifers; however, in another important respect, biological diversity is very different from these other resources. It is similar to these other non-renewables in the sense that it is a one-off endowment from nature to the earth, in that it cannot be replaced on any timescale that relevant to humanity. However, it is distinct from these other resources because it is impossible, by definition, to substitute human innovations for this resource. That is, biological diversity is distinguishable from most other natural endowments by reason of the fact that it is valuable primarily by virtue of its 'naturalness'. It is not possible to substitute human-synthesised inputs or processes for the important characteristics of biodiversity, precisely because their importance derives from the nature of the evolutionary process that generated them. There is no human-synthesisable substitute for the half-billion years' experience that biodiversity represents. Biodiversity management concerns the management of the unique characteristics of this one-time endowment from the evolutionary process.

1.4   THE GLOBAL CONVERSION PROCESS: HUMAN DEVELOPMENT AND DIVERSITY DEPLETION

Although the retention of this evolutionary endowment renders important benefits to all life forms and thus to human societies as well, it is clear that the depletion of diversity has also generated important benefits for human societies. Human societies have been expanding and developing for many centuries through a process closely linked to biodiversity depletion. This is because one of the fundamental avenues to human development has been the conversion of the naturally-existing forms of assets to other forms more highly valued by human societies (Solow, 1974a). This trade-off, between the benefits and opportunity costs of conversion, constitutes the fundamental problem of biodiversity management.

This is because conversion has taken a special form which necessarily implies diversity depletion. Over the past ten thousand years, human societies converting their resources have done so by replacing the naturally-existing slate of species with a selection from the small menu of 'specialised species'. These are the domesticated and cultivated varieties that have been developed for use in agriculture, and are now substituted for the resident diversity worldwide.

The *global conversion process* is the observable result from the sequential application of this process of the replacement of the diverse with the specialised, in country after country across the face of the earth. The nature of this conversion process implies two direct consequences: human societies (and their associated species) have advanced via this conversion strategy, while all other species have been in relative decline.

It is important to sort out the line of argument concerning causation in biodiversity decline. First, the fundamental source of these related phenomena is the human drive for resources and fitness. Second, one of the basic strategies used by the human species in this pursuit is conversion between asset forms. Third, conversion has taken a very special form in the context of the biosphere: the re-allocation of base resources toward a very small selection of species (the domesticated and cultivated species). Fourth, this re-allocation of resources has greatly benefited human societies, while simultaneously reducing the resources available to the vast majority of species on earth (indirectly resulting in their decline and ultimate extinction).

Therefore, the development process has been closely linked with diversity decline over the past ten thousand years. Human advances have been sourced in the systematic reallocation of resources between other species. In addition, to date much of this development gain has been squandered on human niche expansion. Over the past ten thousand years, the human population has expanded from about ten million to approaching ten billion individuals.

One important point to note is that in this framework human population expansion is not a *cause* of biodiversity decline, but its linked outcome. Biodiversity decline is caused by the process of conversion, which has the twin consequences of human advancement and diversity decline. In the past, human advancement has often equated with human population expansion (increased relative 'fitness'). However, this need not necessarily be the case. At base it is the human drive for *wealth or fitness* that motivates conversion; diversity decline as an instrument may be used to achieve either of these aims, but has been used primarily for the latter purpose.

The costs of conversions are less obvious, but equally inherent in this process. Although conversion does not necessarily imply the expansion of the human niche, it clearly does imply the contraction of others. This is because the conversion process has taken the particular form of the replacement of the 'diverse' with a selection from the menu of the 'specialised'. This approach to conversion also has twin outcomes: it expands the niche of a small selection of (cultivated and domesticated) species, while contracting virtually all others. This has guaranteed the loss of a large portion of the world's natural variety, when used as a strategy for global development.

The loss of any single species is difficult to value (hence the ongoing debate concerning the 'sign of option value'). However, there is no ambiguity concerning the meaningfulness of the loss of large swathes of the world's diversity. The homogenisation of the biosphere implies the decline of evolutionarily-supplied diversity, a unique and non-renewable resource. It is the loss of this product, rather than the costliness of the loss of individual species, that is the opportunity cost implicit within the conversion process.

Therefore, the diffusion of the global conversion process has created the observed link between the outcomes of diversity decline and human development. However, there probably is no real necessity of a link between the two. Even assuming that human development requires uniformly 'high' population densities across the world (such as presently exist only in Europe and parts of Asia), this still does not require that the life forms on earth be homogenised in the fashion that has occurred in those places. It is possible for human societies to develop alternative development paths that build on the existing natural base of diversity rather than commencing with conversion to the uniform, specialised resources.

It is this option – *of development compatible with diversity* – that is the focus of this enquiry. Other authors believe that it is only possible to sustain global resources through a policy of non-development; they advocate a pursuit of the policy of the 'steady-state' (e.g. Daly, 1992). The past twenty years have seen the first questioning of the fundamental human pursuits of wealth and fitness, in the context of the debate concerning the 'Limits to Growth'. (Meadows *et al.*, 1972). I have no quarrels with this policy in theory, but I view it as either infeasible or highly objectionable in practice.

A policy of the 'steady-state' would either require all human societies to disavow further growth and development, or it would require that only some societies pursue further wealth and fitness while others maintain or contract. The former option is not feasible, simply because there are many human societies that still exist in a poverty-stricken state. Most of these states would be unwilling to make a commitment to steady-state policies. In addition, it would be unfair to commit some societies to perpetual poverty while others live in great surplus.

The latter option (the selective assumption of development constraints) depends entirely on the identity of the states assuming the constraints. As currently practised, this option is unfair and highly objectionable. It gives rise to allegations of 'environmental imperialism', precisely because these development constraints are usually proposed for imposition upon the most undeveloped states. For example, proposals that states such as Brazil, Zaire and Indonesia should not engage in deforestation and conversion are highly objectionable when they come from the already-converted states of the North.

The only equitable alternative is to pursue *alternative development paths* in the unconverted states by means of investments received from the developed (i.e. to make such investments a method for 'evening out' global asymmetries in development). It is only by way of clear inducements that the North will be caused to invest in a manner that will even-out development. This is the economic role of these unique natural resources such as biodiversity; they provide the foundation on which the unconverted states may stake their claim to a fair share of the global product. The converted states have had the ability to claim a disproportionate share of that product, by virtue of the relative uniqueness of the converted forms of assets (Krugman, 1979). However, with an ever greater share of the world's societies undertaking the same conversions, the truly unique assets now lie in those states that retain the unique endowments of evolutionary product. It is the recognition of this monopoly over this important asset that will cause the converted states to invest significant funds (and thus assume their own development constraints) for the benefit of the unconverted states. However, such investments will only occur when the developing country commits itself to an 'alternative development path', i.e. one consistent with conservation rather than conversion of diverse natural resources.

Therefore, the policy pathway advocated here is for the unconverted states to pursue development in a manner that is consistent with the environmental resource. This will allow development to continue in those parts of the world that are still 'catching up, while simultaneously inducing the converted states to invest in that development process. This strategy recognises biodiversity as a global resource for its universal flow of benefits, but focuses on its character as an exclusive 'national resource' for use as a mechanism for inducing international transfers. This is the policy pathway selected, for reasons of feasibility, sustainability and equity.

1.5 THE REGULATION OF BIODIVERSITY: MANAGING THE GLOBAL CONVERSION PROCESS

The management of extinction (and hence of biodiversity) equates with the management of the global conversion process. The assumed objective is to halt these conversions at least at that point where the human welfare losses from the narrowing of diversity exceed the welfare gains from conversion. This approach indicates a minimum 'baseline' level of biodiversity that should be conserved.

This book argues that the optimal management of biological diversity equates with the channelling of the values of diverse resources from the global community to the suppliers of those benefits, i.e. to those states in the South harbouring significant diverse resources. In that fashion the optimal policy for biodiversity conservation is effected – through a policy geared to harnessing the value of diversity in order to halt conversions – while the burden of diversity conservation is compensated – through the instrument of channelling the value of diversity to its suppliers.

In addition, this manner of diversity management equates with the creation of alternative development paths for states with large areas of remaining wilderness. Development may be made compatible with the retention of diversity; that is, it is not a universal precondition to development that the naturally-existing slate of biological resources be replaced with a pre-selected slate. Investing in institutions for the appropriation and transfer of the values of biodiversity equates with investment in the creation of these alternative development paths.

For this approach to be effective in conserving biodiversity, however, it is necessary to characterise the nature of the extinction process, as it currently exists. Although a wholly natural and life-sustaining process ten thousand years ago, extinction is now driven largely by social forces (i.e. the conversion process). Conversion can be effected via a number of different routes: direct mining for rents (affecting, e.g. many slow-growing tropical forests); indirect conversion to other land uses (affecting, e.g. many virtually unknown plants and insects); and/or the withholding of the ancillary resources required for the management of species exploitation (affecting, e.g. many of the large land mammals – elephants and rhinos). Each form that conversion may take is an equally effective potential avenue to extinction.

Despite the multiplicity in the avenues to extinction, each is based in the same fundamental source: a societal-level determination of investment unworthiness (i.e. the conversion decision). If a society views a species as sufficiently investment-worthy, then stocks of that species will be maintained (through adequate levels of investment in it and its ancillary resources). If the species is not viewed as relatively investment-worthy, then it will suffer disinvestment (via one of the routes described above), and ultimately extinction. It is through the process of investment and disinvestment (in species and their ancillary resources) that conversion operates.

Therefore, this book argues that the 'host state' must perceive diversity as investment-worthy, if its citizens are going to take actions to conserve it. Of course, this implies a fairly innocuous vision of the relationship between individuals, society and the state. Specifically, it assumes that the structure of the state is responsive to the needs of its individual citizens, and especially their joint interests. This implies that societal objectives will be inherent in state decision-making, and that interventions within the state's decision-making framework will impact directly upon the ways in which individual decisions are taken in that society.

This set of limiting assumptions necessarily excludes many other potential sources of conflict and inefficiency in resource management; however, the object here is to demonstrate that, even should host states be responsive to the needs of their citizens, this would not imply the existence of resource management and investment policies that are first-best from the global perspective. Environmental problems are inherent in the division of the global environment into many separate states, as much so as they are inherent in the division of a society into many inter-dependent individuals. Although many other specific causes may be operative in particular cases of biodiversity destruction, these 'inter-national externalities' are the globally applicable and systematic reason for diversity decline. It is role of the 'global community' to intervene within the state's decision-making framework in order to make diversity investment worthy, in order to correct for these externalities.

This book also argues that some of the existing schemes for the international regulation of endangered species are ill-conceived, because they are based upon less fundamental explanations of extinction. In essence, international policies concerning species decline usually focus on the *proximate* causes of their destruction, such as poaching or unmanaged exploitation, without asking why such phenomena occur. When this is the received basis for policymaking, the indicated optimal policies can be the opposite of those indicated above, i.e. the advocated policies are often addressed to the elimination of all incentives to invest in the endangered species. For example, much of the previous economic literature on the regulation of extinction has argued for policies of 'bans' on species exploitation and the withdrawal of consumer markets (Clark, 1973; Spence, 1975). This approach has been incorporated into the existing international scheme for the regulation of endangered species: the Convention on International Trade in Endangered Species (CITES).

A more fundamental approach to endangered species policy recognises that the proximate causes of species decline (poaching, unmanaged exploitation, etc.) are equally the effects of the more fundamental causes of endangerment, i.e. the failure to acquire a sufficient proportion of the value of the diverse resource to warrant the allocation of resources (lands, management, protection, etc.) to the species. A biodiversity policy based on the arguments developed within this book would imply the need for substantial reforms to the existing regulatory framework.

Specifically, it points to the need to create a 'global premium' that is to be conferred upon host states investing in their diverse resources. The premium performs the function of rewarding investments in diverse resources for the non-appropriable services that they render. It also opens the door to the pursuit of these alternative pathways to development, so that states with diverse resources are able to see the advantages to development without the necessity of conversion.

1.6 THE INTERNATIONAL REGULATION OF EXTINCTION: INSTITUTION-BUILDING FOR BIODIVERSITY

The solution to the environmental problem of biodiversity losses is easily stated in the abstract; it is to compensate the supplier-states for the stock-related services that they render. The more difficult problem by far is the creation and implementation of the institution or institutions that will perform this task.

To a large extent, problem and solution are side-effects of a human world with many national governments. The impacts of state by state conversions are non-internalised on account of multinationality. The difficulties of reinternalising the costliness of these conversions is again a function of this global decentralisation. We now live in an era when the global costliness of a world with many governments is becoming apparent, even in the absence of direct conflicts.

The international regulation of extinction requires the creation of an international institution for transferring the values of diversity from consumer-states to supplier-states. It is an attempt to provide a transnational mechanism to compensate for services that are known as 'public goods' in the national context. This institutional problem is at the core of the environmental problem of biological diversity.

Specifically, this book argues that there is a global community of interest derived from the fact that decentralised (multinational) decision-making regarding conversion will lead to local outcomes that diverge from global optima. The international regulation of extinction consists of investments by this 'global community' in international institutions whose purpose is to generate flows of benefits to host states in such a manner that state-level decision-making results in local outcomes that more closely approximate the global optima.

Much of this book addresses the particular characteristics such an institution must have in order to be effective. For example, an effective international institution must provide some manner of assurance of future benefits if it is to impact upon the investment decisions of host states. This is because investment decisions are decisions regarding assets and the anticipated flows regarding them; a state will only deviate from its perceived first-best investment path if the present value of the entire flow of future net benefits from such an alteration would appear to warrant it. Therefore, in order to have a permanent impact on decision-making concerning the selection of development paths, it is necessary to make an impact on the perceived benefits from alternative pathways at all points in time.

This indicates that international institution-building should be directed to the permanent alteration of the terms of trade between specialised and diverse investments. At present the terms of trade are biased toward the specialised forms of resources on account of the relative rates of appropriability and rent capture. International institutions could create enhanced benefit flows to investments in diversity by one of three potential routes: enhanced rent capture ('international wildlife trade regulation'); enhanced appropriability ('international intellectual property rights regimes'); or direct subsidies to non-conversion ('international franchise agreements'). The former regimes operate indirectly, by enhancing the returns to diversity-related investments. The latter is direct intervention; it provides a stream of benefits in return for the act of non-conversion (also known as a 'transferable development right').

The argument here is that international institution-building must focus instead on the creation of *alternative pathways to development*, not the acquisition of a society's right to develop. The distinction is fine, but crucial. First, a focus on alternative pathways emphasises the importance of creating a constructive outlet for the fundamental human drive for advancement; purchasing rights to specific routes for advancement leaves this process undirected. When undirected, the purchase of development rights becomes a 'policing' agreement: enforcing low-intensity use agreements on behalf of an absentee landlord. A more constructive approach harnesses the human drive for development rather than combating it.

Secondly, the particular pathway emphasised – *development from diversity* – implies an assumption about the specific nature of global environmental problems. The assumption is that these problems derive less from the general scale of human activities than from the extreme scale of very specific human activities. Environmental systems are able to withstand higher levels of many diverse activities (drawing upon many different resources and systems) than lower levels of human activity concentrated on a single resource sector or system. It is the uniformity of development as much as its scale that depletes environmental resources. Investing in diverse pathways for development is therefore synonymous with investment in a broad range of these resources and systems. International institution-building for biodiversity should have these considerations as its general objective.

1.7 THE INTERNATIONAL REGULATION OF EXTINCTION: PROSPECTS AND PROSPECTUS

In the summer of 1992 the global community met in Rio de Janeiro and discussed most of the issues that are the subjects of this book. One of the end results of the United Nations Conference on Environment and Development was the adoption of a text for an international convention on the conservation of biological diversity.

The language and legal content of this convention is, at present, exceedingly vague, representing unspecified undertakings on the part of the developed world to 'transfer technology' and to 'compensate incremental costliness' in return for unformed commitments on the part of the developing world to 'conserve diverse resources' and to 'manage biological diversity'. This agreement is of the nature of a 'framework convention'. It identifies the issues being discussed and the parties to the negotiations, but it goes only a very short distance in the direction of a solution.

It is the thesis of this book that the major difficulty facing the parties to these ongoing negotiations is an understanding of the problem with which they are dealing. Only after a common conceptual framework for viewing the problem is agreed can the nature of a solution be identified. It is the purpose of this book to make a first attempt at these tasks. First, it attempts to define and defend a scientific framework within which the global problem of biodiversity might be approached. Then, with the nature of the problem identified, it attempts to specify the nature of the solution indicated by the form of the problem.

Therefore, the purpose of this volume is to provide a substantive basis for the development of an effective international management regime for the conservation of biological diversity. That is, it is an attempt to 'fill in the blanks' of the biodiversity convention.

This task is accomplished in the context of ten chapters, in the following fashion. First, Chapter 2 provides the reader with some concrete illustrations of some of the problems and policies with which global biodiversity is concerned. It provides some background material in order to substantiate the 'stylised facts' on which the remainder of the analysis is built. It also previews many of the fundamental concepts developed in more detail in subsequent chapters.

Chapter 3 is a review of the economics literature of extinction, and an introduction to a more fundamental approach to the issue. This chapter demonstrates that the existing literature focuses on the proximate, rather than the more *fundamental causes of extinction*, and therefore provides misleading conclusions for extinction policies. It argues for an overhaul of both extinction theory and policy.

Chapter 4 is an attempt to provide a new theory of human-sourced extinctions. It commences with a brief review of some of the biological literature on extinction. It then draws the distinction between the natural process of extinction and the current, human-sourced, extinction problem. And, most importantly, it demonstrates that the prior explanations of human-sourced extinctions, mismanagement and overexploitation, cannot explain the nature of the current extinction process. Extinctions, and diverse species decline, have been too closely linked with the expansion of the human niche to have been the result of mismanagement. Rather, it is the case that human 'management', i.e. human reshaping, of the biosphere for the purposes of the advancement of human wealth and fitness is the base explanation for the decline of diverse species. Therefore, Chapter 4 outlines the nature of the *global conversion process*, and argues that this is the fundamental explanation for diversity depletion.

Chapter 5 links the previous two chapters with the following two. It provides a theoretical analysis of state decision-making regarding the management of its natural resources. Since all natural resources are potentially 'common' resources, it is necessary to discern why the state decides to allocate different forms of management regimes to different resources. Chapter 5 demonstrates that one of the investment decisions that a state must make is the extent to which it will invest in the development of institutions to manage its various natural resources, and that when investments are not justifiable then specific resources go unmanaged. When biological resources are unmanaged, overexploitation is assured. Therefore, Chapter 5 demonstrates that *overexploitation is underinvestment*, i.e. even species declines that match up with the previous ('open access overexploitation') theories of extinction are determined by more fundamental forces. In this way, this chapter demonstrates in specifics what Chapters 3 and 4 argued in generalities.

More importantly, Chapter 5 also provides the underlying model of the state's decision-making framework which establishes the basis for the ensuing four chapters on the international regulation of biological diversity. That is, since terrestrial resources are national resources, the nature of international regulation must necessarily be one of intervening within national decision-making processes. Chapter 5 (specifically, equation (5.15)) establishes the basic model of state-level decision-making upon which the remainder of the book is built.

Chapter 6 establishes the general objective of *the optimal policy for biodiversity conservation*. It shows that the fundamental objective of biodiversity policy is to manage the global conversion process in order to maximise global benefits. Then, it demonstrates the nature of those global benefits, specifically the nature of the value of evolutionary outputs. This chapter studies the trade-off between the benefits received from further conversions and the global costliness implicit in the further decline of diversity.

Chapter 7 analyses *the general nature of the international agreement required for biodiversity conservation*. Biodiversity conservation requires intervention within the host state's decision-making framework, and this institution-building is of a very specific form. Specifically, it requires ongoing investments by the global community in order to alter the present value of the flow of benefits from further conversions. This chapter investigates how this objective might be accomplished via the instrument of direct subsidies within the context of international agreements.

An alternative to the alteration of the net benefits from conversion is intervention to enhance the benefits received from investments in diverse resources. This implies institution-building to enhance either rent capture or appropriability with regard to diverse resource flows. Chapter 8 provides a case study in the difficulties in implementing an international institution for this object. It does so in the context of analysing the use of another instrument, trade regulation, for the conferral of enhanced benefits upon diversity suppliers. The international regulation of wildlife trade could be translated into a mechanism for rent appropriation, thus making the production of diversity a relatively more attractive pathway for development.

Since international institutions have existed in the area of wildlife trade regulation for over twenty years, this chapter provides the opportunity for examining the difficulties involved in the actual implementation and evolution of an international instrument for diversity conservation. It demonstrates the maze of legal and institutional complexities that arise in the implementation of a simple concept such as rent enhancement.

Chapter 9 examines the most controversial element in the diversity regulator's arsenal: intellectual property rights. It analyses the basic character of this instrument, used for the purpose of providing incentives for informational investment. It then examines its applicability in the conservation context. The idea of intellectual property right regimes is to render the inappropriable value of certain forms of services appropriable, by state sequestration of a product market monopoly. The range of societal objectives to which this instrument has been applied has to date been fairly narrow, essentially consisting of those forms of services that result from investments in human capital. Chapter 9 argues for the extension of the same regime for the protection of investments in natural capital.

Chapter 10 concludes with a summary of existing policies relating to diversity conservation, together with a set of proposals for reforms in these areas. In this chapter the 'blanks' of the framework Biodiversity Convention are 'filled-in' with calls for protocols on trade regulation, intellectual property rights and international franchise agreements.

1.8 CONCLUSION

The international regulation of extinction requires international institution-building for the purpose of creating alternative pathways to development. To the present, the incentives for international institution-building have, of course, been based around already existing industries and the development path that they imply. The problem with this approach to international institutions is that this dictates the use of the same development paths by those states considering their initial steps toward development as were used by those states which have already developed. In this way, existing institutions generate a progression toward a uniform, non-diverse world.

The difficulty here is that the depletion of diversity has its own costliness. As each successive state embarks upon precisely the same pathway to development, commencing with the conversion of its biological resources, the costliness of this conversion decision is increasing. The sequential selection of the same development path, and the consequent concentration on the same natural resources and ecological systems, creates unnecessary pressures on these sectors. As the last states to embark on this development path convert the remainders of diversity, the global costliness of their decisions could be virtually limitless. Global environmental problems derive as much from the sameness of human activities as from their scale.

International institution-building must be refocused in order to provide development options suited to the initial conditions of these last, developing countries. That is, it is necessary to invest in institutions that are suited only to the encouragement of development from diversity, as much as it is to invest in institutions that are suited only to conversion. This is the first step to conserving diversity on earth: institutional, economical and biological.

2 Global Biodiversity: Some Background and a Preview

2.1 INTRODUCTION

The purpose of this chapter is to provide some background information on the processes, problems and policies associated with global biodiversity decline. In doing so, it will also provide a preview of some of the theory that will be developed in more detail in later chapters. Section 2.2 presents an outline of the nature of the global conversion process, and some examples of its 'twinned outcomes' of diversity decline and societal development. Section 2.3 presents a depiction of the global biodiversity problem; it is shown that this problem results from the existing decentralised approach to the global conversion process. It is necessary to construct an international system of regulation for a global resource that will internalise the externalities flowing from accumulating conversions. Section 2.4 provides an outline of the optimal policy for managing the global conversion process. It shows that an optimal policy will operate through internalising the values of diversity, in order to counterbalance the values derived from conversions, and it gives examples of the nature of these values.

2.2 THE GLOBAL CONVERSION PROCESS: FORCES DRIVING THE GLOBAL EXTINCTION PROCESS

Extinctions now result from the failure of human societies to invest in certain species. The fundamental source of the global extinction process is the human reconstruction of the global portfolio of biological resources on which they rely, from which most biological resources (the so-called 'diverse' resources) are being excluded. When a given species is not selected for inclusion within this portfolio, then it is subjected to the forces for disinvestment that lead to its decline.

This process is termed the *global conversion process*, and it has existed on earth for nearly ten thousand years. Since that time, human societies have applied new technologies which are imbedded in a small slate of specific species (the domesticated and cultivated varieties). It is the extension of this technological change, for the advancement of human wealth and fitness, that is the fundamental force for extinction in operation today.

2.2.1   Distinguishing Fundamental and Consequential Causation in Extinction

Although the fundamental force generating biodiversity losses is the omission of many species from this global portfolio, the proximate cause of a given species' decline may be one of several, consequential types. Exclusion implies disinvestment in the species, but the nature of the disinvestment process depends upon the nature of the species and its relationship to human society.

First, the species may be seen as valuable, but not as an asset (i.e. not as a resource with potential for growth); then it will be perceived to be optimal to harvest the entire stock of the species in order to invest the return in another, more productive asset. This is direct 'stock disinvestment' for conversion to another, preferred asset. An example of this form of disinvestment would probably be much of the deforestation that has occurred over the past few hundred years.

Second, a species may not be perceived as valuable, as an asset or otherwise, and thus it will not be worthy of substantial investments in the ancillary resources required for its maintenance. The most obvious requirement for species survival is an allocation of 'base resources': the natural habitat that it requires for its sustenance. When a species is not selected for society's asset portfolio, then the allocation of base resources which it depends upon for survival is usually diverted to the use of other, chosen assets. An example of this form of disinvestment would probably be the projected losses of general diversity (unclassified insects, plants, etc.) set to occur in the tropical zones over the next century.

Third, the species may be seen as valuable, but not valuable enough to warrant the allocation of the resources required for a managed disinvestment programme. Then the species will again incur disinvestment, but on this occasion the disinvestment will occur by reason of the lack of investment in the management of access to the species or its habitat. This results in what is commonly called 'open access overexploitation'. An example of this form of disinvestment would probably be the decline of many of the large land mammal species, such as the rhinoceros and the elephant.

Although these three routes to disinvestment are the visible layer of causation in the extinction process, the more fundamental forces lie beneath these within the decision-making processes regarding society's asset portfolio. The important issues concern the nature of this selection process, and the forces driving it toward global homogenisation.

The process of the selection of assets for society's portfolio is an economic decision, determined by forces that shape the perceived relative advantageousness of different assets. These fundamental forces are reshaping the biosphere by creating the impression that *non-diverse resources* are relatively advantageous assets as compared with *diverse resources*. Sections 2.2.2 to 2.2.4 discuss this distinction between diverse and non-diverse, and the forces that operate to cause human societies uniformly to select the latter over the former.

2.2.2 The Conversion Process and the Biosphere

In composing their optimal portfolio of assets, human societies must choose between the many different forms of assets in which they might hold their aggregate wealth. If the optimal portfolio determined by this process is different from the initial, then there will be incentives to convert some existing assets to other forms. This implies disinvestment in some naturally-occurring assets, and investment in some other human-preferred ones.

This incentive to disinvest in the naturally-occurring stock of assets in order to invest in the human-selected is the 'conversion process'. The natural form of any resource is necessarily competitive with other forms in which humans might hold these same assets. This derives from the fact that humans now have the capability to choose whether to hold terrestrial natural resources in their original form, or to substitute another (Solow, 1974a).

This force for conversion applies even to biological resources. For example, a given hectare of land, which is originally growing diverse native grasses, may be converted directly to another grassy plant form such as wheat, because of the enhanced productivity of this resource. Alternatively (and less directly), a tropical forest may be logged and sold with the funds then invested in other national assets (such as education), resulting in the conversion of the natural asset to, for instance, human capital.

Economic development in human societies derives in part from the substitution of the more productive assets for the less productive, i.e. from the conversion process. However, the application of this economic process on a *global basis* is one of the primary forces contributing to diversity losses. The initial, local conversions of natural resources had little impact on the global portfolio of assets, but the occurrence of thousands and millions of these discrete conversions has generated a phenomenon of worldwide importance.

In effect, the global conversion process may be conceived of as the diffusion of the idea of asset conversion across the globe, from state to

Table 2.1   Countries with the greatest 'species-richness'

Mammals	Birds	Reptiles
Indonesia (515)	Colombia (1721)	Mexico (717)
Mexico (449)	Peru (1701)	Australia (686)
Brazil (428)	Brazil (1622)	Indonesia (600)
Zaire (409)	Indonesia (1519)	India (383)
China (394)	Ecuador (1447)	Colombia (383)
Peru (361)	Venezuela (1275)	Ecuador (345)
Colombia (359)	Bolivia (1250)	Peru (297)
India (350)	India (1200)	Malaysia (294)
Uganda (311)	Malaysia (1200)	Thailand (282)
Tanzania (310)	China (1195)	Papua NG (282)

Source: McNeely *et al.* (1990). *Conserving the World's Biological Diversity.* International Union for the Conservation of Nature: Gland, Switzerland.

state. On its arrival, the state on the existing margin has the opportunity to restructure its national investment portfolio, and hence its natural resources. The global biodiversity problem comes to our attention at this point in time because these processes of conversion are now working their way towards the last refugia on earth. The majority of the world's remaining species reside in a small number of the world's states (Table 2.1). These are the same states that have been the last to have substantial parts of their territories remaining unconverted.

Asset conversion that has occurred for millennia on a local and regional scale has now aggregated to become a force at the global level. At base, this restructuring of the global portfolio of biological assets is driven by the desire for human development gains obtained from the conversion of assets to more productive forms. However, as this basic strategy for human development reaches the final refugia of many of the world's species, it is projected that a cataclysmic 'mass extinction' of species may result (Ehrlich and Ehrlich, 1981; Lovejoy, 1980).

Therefore, at the very base of the biodiversity problem is the capability of humans to change the nature of the biosphere from its natural to a human-preferred form. The gains from conversion, and the drive for human development and fitness, have been causing the restructuring of the biosphere on a regional basis for several millennia. Now, with the diffusion of this strategy to the final terrestrial frontiers, conversion of the biosphere seems set to occur on a global basis.

2.2.3 The Nature of the Global Conversion Process

Reconstruction of the portfolio of biological assets on a global basis is a powerful force, capable of reshaping the whole of the earth's biosphere. However, it is not in itself sufficient to explain the potential for a mass extinction. For this, an explanation must be found that will generate not only an expected reshaping of the global portfolio of natural assets, but also a narrowing of that portfolio.

Conversion as an economic force explains only why it is the case that the natural slate of biological resources might be replaced by another on any given parcel of land, depending upon relative productivities, but it does not explain why a small number of species would replace millions across the whole face of the earth. That is, this force implies conversion but not necessarily homogenisation. In order to explain the global losses of biodiversity, i.e. *a narrowing of the global portfolio*, it is necessary to identify the nature of the force that would generate this homogenisation of the global biosphere.

It is unlikely that a wholly natural process would drive the world toward substantially less diversity. This would require the evolution of both biological generalists (species with superior productivity across many niches) and uniform human tastes (across the globe). In fact, the current drive toward uniformity is contrary to the very idea of evolutionary fitness. Fitness implies competitive adaptation to the specific contours of a certain niche. The evolutionary process generates species that are well-adapted to their own specific niches through a process of niche refinement; that is, a surviving species represents a 'good fit' to its own niche (Eltringham, 1984). The prospect for human societies happening upon a generalist species capable of outcompeting resident species in their own niches is very small.

It is equally unlikely that human tastes are so uniform as to demand the homogenisation of biological resources. Communities 'co-evolve' in order to better fit with the system in which they participate. It would be expected that the preferences of predators would be determined generally by their available prey species. In fact, there is ample evidence to support the expectation that human communities would prefer to consume the resources upon which they traditionally depended (Swanson, 1990b).

This indicates that the depletion of diversity is not a natural phenomenon; rather, it is a socio-economic one. There are good reasons to believe that prevailing methods of production are biased against the maintenance of a wide range of diversity. The idea of agriculture, that originated about ten thousand years ago in the Near East, was centred on the idea of creating species-specific technologies. This implied the inclusion of two new important factors of production in the production of biological goods: species-specific capital goods and species-specific knowledge.

In terms of biological resources, the capital goods applied in production are the chemicals, machinery, and other tools of agriculture. These capital goods usually do not enhance the photosynthetic productivity of the biosphere; rather, they increase its productivity by means of the mass production of large quantities of a homogeneous output from much-reduced inputs from other factors, e.g. labour.

The productivity gains in agriculture go hand-in-hand with diversity losses; in fact, they are often derived from the reductions in diversity. For example, farm machinery is developed to work in fields that are planted uniformly in a single crop. Chemicals are fine-tuned to eliminate all competitors of a single species. The fields themselves are 'cleared', for the introduction of the machinery and chemicals of the production process. These capital goods are effective precisely because of the homogeneous environment within which they operate, and they create incentives for conversion by reason of their effectiveness.

At present, this process of conversion is working its way across the developing world, having completed its journey through the developed world. The frontier is discernible by reference to the relative rates of conversion and capital good accumulation. For example, the number of tractors in Africa increased by 29 per cent over the past ten years; they increased by 82 per cent in South America; and by 128 per cent in Asia. During the same period the number of tractors decreased by 4 per cent in North America (World Resources Institute, 1990). Similarly, the amount of land dedicated to specialised agricultural production increased by more than 35 per cent throughout the developing world since 1960, while declining slightly in the developed (Repetto and Gillis, 1988). It is the extension of this previously successful strategy for development to the four corners of the earth that is at the base of the concerns about what is presently happening to the biosphere.

2.2.4 Global Non-convexities: Convergence on Specialised Varieties of Species

The other important factor introduced into the production of biological goods was species-specific learning. With more experience with a particular species, it was possible to become even more efficient in its production (by reason of increased understanding of its biological nature, as well as intervention to determine the same). This information became another crucial factor for agricultural production, but it existed only in one form – embedded in the received forms of the domesticated and cultivated varieties.

Agriculture originated approximately ten thousand years ago in the Near East. It consisted of a set of ideas, a set of tools, and a set of selected species. At that time and in that locale, each of these selections was locally optimal. However, the set of ideas–technology–species was transported out of that region as a single unit, as the continuing investments in this combination caused the ideas and tools to become embedded in the chosen species.

A non-convexity was introduced within the decision-making process, by reason of the non-rival nature of the information embedded in the specialised species (that would be costly to produce for any diverse species). There is the essential difference between the specialised (domesticated) species and the diverse (wildlife) species. For one group, an information set is publicly available as an input into their production; for the other, it is necessary to construct that same information.

The global conversion process has consisted of the extension of these chosen species' ranges. As a consequence, much of the face of the earth has been reshaped in order to suit these few species and the tools used in their production. It is the diffusion of this 'bundle' of ideas–tools–species that is at the base of the biodiversity problem.

Therefore, it is not simply the globalisation of the strategy of asset conversion that is determining the global portfolio, it is also the special way in which conversion occurs under agriculture. It is the perceived gain from the substitution of the specialised biological resources for the diverse that is generating an ever more narrow portfolio. It is this force, now acting globally, that is shaping the incentives for investment, and hence extinction.

This is a form of dynamic externality in operation with regard to decision-making concerning the production of biological goods. That is, subsequent choices regarding conversions are being determined by previous ones. In the context of the biosphere, this non-convexity is creating a 'natural monopoly' for a small number of species. The biosphere is converging upon this small, select group of specialised species as the sole providers of biological goods to human societies.

This is evidenced by the increasingly narrow roster of species which meets all of the needs of humankind. Of the thousands of plant species that are deemed edible and adequate substitutes for human consumption, there are now only twenty which produce the vast majority of the world's food (Plotkin, 1988). In fact, the four big carbohydrate crops (wheat, maize, rice and potatoes) feed more people than the next twenty-six crops together (Witt, 1985). The same applies with regard to protein sources. The Production Yearbook of the Food and Agricultural Organization lists only a handful of domesticated species (sheep, goats, cattle, pigs, etc.) which supply nearly all of the terrestrial-sourced protein for the vast majority of humans. The number of domesticated cattle on the globe (currently over 1.2 billion or one for every four humans) continues to increase, while the numbers of almost all other species continue in decline (World Resources Institute, 1990).

The same is true with regard to variety within a species. Not only are human societies becoming more reliant upon a narrower range of species, they are also becoming reliant upon specific varieties of these species. Specialisation works beyond the species level of genetic convergence to produce a technically calibrated uniform biological asset.

The global diffusion of specialised species is demonstrated in Tables 2.2 and 2.3. The first table provides a static portrait of the progress of this technological change in the period 1978–81. It shows that some developing countries had already embraced this strategy of specialisation (e.g. the Philippines with 78 per cent of their rice production converted) while others were only just initiating the process (e.g. Thailand with only 9 per cent of the same).

Table 2.2   Area devoted to modern rice varieties
(seven Asian countries 1978–1981)

Country	Year	1000 ha	% of rice area
Bangladesh	1981	2 325	22
India	1980	18 495	47
Sri Lanka	1980	612	71
Burma	1980	1 502	29
Indonesia	1980	5 416	60
Philippines	1980	2 710	78
Thailand	1979	800	9

Source: Hazell, P. B. R. (1985). 'The Impact of the Green Revolution and the Prospects for the Future'. *Food Reviews International*, Vol. 1, no. 1 (1985).

Table 2.3 shows the progress of this process within individual states. In those states that initiated modern agricultural specialisation (e.g. the USA), food production is now almost entirely specialised (the majority of food production involving only a few varieties of a small number of species). In the states adopting the strategy more recently, this 'scoping in' process has reduced the number of varieties in production from thousands to a few in a small amount of time.

The fundamental nature of the biodiversity problem derives from this particular form of the global restructuring of the biological asset portfolio. Diverse resources suffer from disinvestment primarily because these few specialised varieties monopolise society's investments in biological assets.

Table 2.3   Examples of genetic uniformity in selected crops

Crop	Country	Number of varieties
Rice	Sri Lanka	from 2000 varieties in 1959 to 5 major varieties today (Rhoades, 1991) 75% of varieties descended from one maternal parent (Hargrove *et al.*, 1988)
Rice	India	from 30 000 varieties to 75% of production from fewer than 10 varieties
Rice	Bangladesh	62% of varieties descended from one maternal parent (Hargrove *et al.*, 1988)
Rice	Indonesia	74% of varieties descended from one maternal parent (Hargrove *et al.*, 1988)
Wheat	USA	50% of crop in 9 varieties (National Academy of Science, 1972)
Potato	USA	75% of crop in 4 varieties (NAS, 1972)
Cotton	USA	50% of crop in 3 varieties (NAS, 1972)
Soybean	USA	50% of crop in 6 varieties (NAS, 1972)

Source: World Conservation Monitoring Centre, 'Valuing Biodiversity', in World Conservation Monitoring Centre, *Global Biological Diversity*, Chapman and Hall: London.

The net effect is the displacement of the former by the latter on a global basis.

2.2.5   The 'Uneven' Nature of Global Conversion: Human Development and Diversity

These conversions from diverse to specialised resources have generated substantial worldwide productivity gains. World cereal production grew at an average annual rate of 2.7 per cent between 1960 and 1983 (Hazell, 1989). For example, the substitution of specialised rice varieties for diverse is estimated to increase yields by 1.0 tonne/hectare on irrigated lands, and by 0.75 tonne/hectare on non-irrigated lands (CIAT, 1981). Although the conversion of lands from diverse to specialised production methods must reduce global diversity, it is apparent that these losses are compensated for, and driven by, development gains.

The economic relationship between conversion and development is demonstrated in part by the state of human development in the 'diversity-rich' states (Table 2.4). Almost without exception, these are some of the poorest nations on earth in terms of human wealth. They range between one and seven per cent of the OECD average per capita income. Although

Table 2.4   GNP per capita in the species-rich states

Country	1988 GNP p.c. ($)	Country	1988 GNP p.c. ($)
Tanzania	160	Papua, NG	810
Zaire	170	Thailand	1000
Uganda	280	Bolivia	1099
Ecuador	284	Colombia	1139
China/India	340	Peru	1300
OECD Average	$17 400		

Source: The World Bank (1990). *World Development Report*. World Bank: Washington, DC.

non-human species are faring relatively well in these countries, the human species is doing comparatively poorly.

From this perspective, the decline of diversity has been closely linked with the human development process. The drive for wealth and fitness has been the fundamental force causing humans to practice conversion. The conversion of biological resources took the form of substituting the specialised species for the diverse, causing diversity to decline. This then generated a gain for that human society, a gain that could be allocated either to increased wealth or fitness. Thus, conversion, and especially conversion to the specialised species, has been a strategy for generating human development gains.

To date, much of the gain achieved from this strategy has been expended on the expansion of the human niche, i.e. increased human 'fitness'. For the human species, a revolution in niche expansion has occurred over the last ten thousand years. Scientists estimate that the introduction of the ideas of agriculture at that time coincided with a 'take-off' in the level of the human population. Since that time, the human population has expanded from approximately ten million to approaching ten billion individuals (see e.g. Boulding, 1981).

Despite the scale of the human population, it remains the method of appropriation that is the gravest threat to diversity. This has been demonstrated in various ecological studies. The ultimate scarce resource, biologically speaking, is known as Net Primary Product (NPP). This is the total biomass generated by the process of photosynthesis on this planet. It is also the total amount of usable solar energy available for the sustenance of all life forms on earth. The expansion of the human niche has resulted in the exclusion of most other species from a substantial part of NPP. Ecological studies show that the human species now appropriates about 40 per cent of this global product (Vitousek, Ehrlich and Ehrlich, and Matson, 1986).

Most importantly, however, the Vitousek study argues that the vast majority (90 per cent) of all human niche appropriation occurs 'indirectly', i.e. for reasons other than direct use. These 'indirect uses' include clearing, burning and otherwise altering lands. This demonstrates that it is human niche appropriation (the denial of other species' use of resources), rather than human niche expansion (the consumption of resources), that is currently endangering most of diversity. The concepts can be unlinked.

In essence, the global process of diversity depletion has been used as an instrument for human societal advancement. Diversity depletion contributes to the human development process through the enrichment of one species by means of the appropriation of the resource bases of millions of others, even though the human species makes direct use of only a small proportion of that which is appropriated.

The conversion process has been working somewhat sequentially across the states of the earth. It is not difficult to ascertain the approximate location of the technological frontier in this context. For example, data on worldwide land use trends document the rates at which conversions of lands to uses in specialised agricultural production have been occurring. Between 1960 and 1980, the developing world in aggregate increased its land area dedicated to standard specialised crops by 37 per cent, while the developed world experienced a small decrease in the same (Repetto and Gillis, 1988). Therefore, deforestation and land use changes continue to occur on a large scale in those countries with natural resources remaining to convert. In other parts of the world, there remains little left for conversion. For example, the amount of 'wilderness' (i.e. 400 sq. km. of unaltered landscape) on the European continent is now virtually zero, versus a global average of approximately 30 per cent (World Resources Institute, 1990). These states of the North are the 'already converted' states; it is only a small selection of the states of the South that retain a significant amount of diverse resources.

At present, the forces for specialised conversions have moved to the boundaries of the last handful of states with substantial amounts of unconverted territory: Brazil (and the other Amazonian states), Zaire, Indonesia and a few others. These states are in a rapid phase of development and conversion, following in the paths of all those states that have gone before. One indicator of the rate of this change is the scale and extent of the land use conversions and deforestation occurring in these countries. Another facet of such conversions is the routing of the productivity gains achieved. Still, these gains are usually routed initially to the expansion of the human niche, and this is indicated by the growth in the human populations on the conversion frontier (Table 2.5).

Table 2.5   Population growth in the species-rich states
(per cent per annum, 1980–90)

State	Growth rate	State	Growth rate
Tanzania	3.1	Papua, NG	2.5
Zaire	3.2	Thailand	1.8
Uganda	2.5	Bolivia	2.5
Ecuador	2.4	Colombia	2.0
China	1.4	Peru	2.3
India	2.1		
OECD Average	0.6		

Source: The World Bank (1992). *World Development Report*. World Bank: Washington, DC.

Therefore, development (human development) is a process that has been driven in part by the process of conversion. This has resulted in a remarkable asymmetry in the world. The states with high 'material wealth' have low 'diversity wealth', and vice versa. The problem of biodiversity stems primarily from the attempts of the remaining, unconverted states to follow this same development path. At present, the margin of the global conversion process rests at the threshold of the last refugia for diverse biological resources.

2.3   THE GLOBAL PROBLEM OF BIOLOGICAL DIVERSITY: REGULATING THE GLOBAL CONVERSION PROCESS

The global biodiversity problem (as a regulatory problem) derives from the fact that this conversion process has been regulated on a globally decentralised basis. That is, each state has been able to make the conversion decision regarding its resources without reference to the potential global effects. This creates an important regulatory problem because the cost – in terms of the value of lost services – of each successive conversion is not the same. The global stocks of biological diversity generate a flow of services to all societies on earth. The first subtractions from global stocks did little to hinder the flow of these services, but the final subtractions from these stocks will render these flows non-existent. As the last refugia for diverse species dwindle, the cost of each successive conversion (in terms of diverse resource services lost) escalates rapidly.

Although it may be threatening the existence of a flow of services from global stocks of biological diversity, the continued depletion of these stocks may nevertheless be to the benefit of the individual society that is undertaking it. This is the nature of the regulatory problem of biodiversity losses – it is a conflict between what is in the interests of the development of the individual country and necessary for the protection of the production system relied upon by the global community. The individual country simply wishes to undertake the conversion process, as have all states that have preceded it in this development process, while the global community wishes to internalise the global costliness of the final conversions to these last, unconverted states.

Therefore, the global policy problem of biodiversity losses involves the management of the global conversion process so as to reach the correct endpoint, taking into consideration the 'global externalities' that individual societies do not. That is, it is necessary to ascertain a 'global stopping rule' that will determine when the marginal conversion is not globally beneficial, and then alter the decision-making framework of the marginal state so that the conversion will not occur.

Figure 2.1 demonstrates the nature of the global conversion process, and the regulatory problem concerning it (or, the global biodiversity problem). It shows that the development process drives society to convert more and

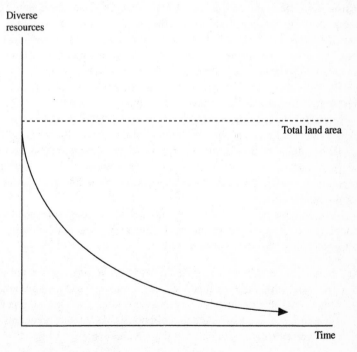

Figure 2.1 The global conversion process

more of its land area to specialised uses over time. Each such conversion confers a gain upon human society – the value of converting between assets – and thus continues to drive the conversion (and development) process. The pertinent questions then become: What forces might halt the conversion process prior to total conversion? What countervailing force is there to offset the effects of the force for specialised conversions?

It is the value of diversity itself that should provide the stopping-point in the global conversion process. That is, with successive conversions, the quantities of lands in specialised production will be increasing while the quantities in diverse resources decline; at some point in this process, the relative values of the two uses might switch, so that the use of the land in diverse resources is preferred. It is the value of biological *diversity* that should arrest the conversion process at its optimal point. The stock of global diversity provides important inputs into the processes of biological production, and it is this value (and not the individual values of the bio-logical materials themselves) that is the essential force to be given effect within the biodiversity regulatory process.

Without intervention, it is very unlikely that this force will be of any effect. As indicated, the main source of benefits from diverse resources lies in their 'stock-related values'. In other words, these are benefits that accrue more to the world at large, rather than to the state hosting them. Such diffuse values will not in general be taken into consideration in state decision-making regarding conversion. If diverse biological resources are systematically undervalued, then they will be too readily converted to their specialised substitutes. This will result in the retention of a quantity of diverse resource stocks that is less than optimal.

Figure 2.2 demonstrates how the non-appropriability of these stock-related values will lead to the mistargeting of the conversion process. That is, this is a diagram illustrating the impact of the conversion process over the very long run, as conversions erode the remaining diverse resource stocks, on the relative values of lands in specialised and diverse biological resources.

This diagram demonstrates that the quantities of lands dedicated to the production of specialised resources in the very long run (allowing all factors to adjust) will be determined by:

(1)	Supply of conversion (S) – This downward-sloping curve represents the *internalised* marginal cost of converting to specialised resources. Each state that decides to convert to specialised resources perceives decreasing costliness on account of choices made by its predecessors, and pre-existing investments in these technologies.

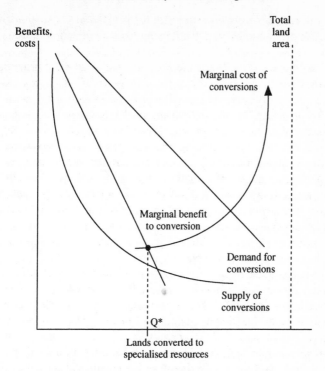

Figure 2.2 Optimal policy regarding conversions

(2)   Demand for conversion (D) – This is the perceived benefit to the
      marginal state from the conversion of its resources (i.e. reshaping its
      portfolio from diversity to specialisation). This benefit is declining
      because there is consumer resistance to the acceptance of specialised
      substitutes for some natural resources. It is also declining because the
      by-products of conversion, i.e. human niche expansion and develop-
      ment, probably yield positive benefits (with consequent population
      growth, urbanisation and industrialisation) but at a declining rate as
      these characteristics become less scarce with additional conversions.

(3)   Marginal cost of conversion (MC) – One of the important marginal
      costs of the conversion of lands from diverse resources to specialised
      is the opportunity cost of forgone diverse resource stocks. These
      stocks are included in MC, but not in S, because these represent the
      'true' rather than the 'supply' costs of conversions.

The optimal quantity of lands dedicated to specialised production is
represented by the point Q*, which is where the marginal piece of land

remains unconverted because the marginal benefits of conversion are no greater than the marginal benefits flowing from its retention in its natural state.

The divergence of the S and the MC curves in Figure 2.2 provides the explanation for the mistargeting of the global conversion process. In this scenario, the supply curve for specialised lands is misperceived, because of the failure to internalise the full costliness of specialised production. The global externalities are not being considered in the supply cost of marginal lands, and as these are increasing with each successive conversion (and especially when the final stocks are endangered), the supply curves deviate from one another more substantially with each conversion. The individual or state making the conversion decision considers only this costliness (within a 'decentralised' regulatory framework), and thus an excessive quantity of specialised lands results (possibly even total conversion).

The global problem of biodiversity is the result of this decentralised approach to the global conversion process. Each state has converted its lands to specialised resource production without consideration of the stock-related costliness of these decisions. Early conversions were able to be undertaken at low global costliness (because S and MC did not diverge significantly when substantial quantities of other stocks remained). However, as the final stages of the conversion process are undertaken, this divergence becomes increasingly severe and ultimately unbounded. The global problem of biodiversity involves the creation of an international regulatory mechanism which will bring this divergence within the decision-making framework of the remaining, unconverted states.

2.4   OPTIMAL BIODIVERSITY POLICY: APPROPRIATING THE VALUES OF GLOBAL BIOLOGICAL DIVERSITY

The divergence between the two supply curves (actual and perceived) in Figure 2.2 represents the value of biological diversity. The area between the two curves represents the total value of biological diversity (ultimately unbounded), while the distance between the two curves at any particular point in the conversion process is the marginal value of biological diversity. That is, the difference between MC and S at a particular point in the conversion process is the value of the marginal stock of diverse resources under consideration for conversion. It is the *marginal value of biological diversity* that should halt the conversion process, but is unable to do so when this value is not appropriable.

The optimal policy for the global problem of biological diversity is to ensure that the states acting as the providers of diverse resource services appropriate this marginal value of their remaining stocks of diverse resources. The implementation of this policy will halt the global conversion process at the optimal point from the global perspective. This means that the understanding and appropriation of the value of *diverse* resources is the key to optimal biodiversity policy.

Diverse resource stocks have always provided value to human societies. Besides their obvious existence values, they also have an important role as inputs into the biological production system, even into the very specialised system that currently exists. Specifically, diverse resource stocks are useful for the insurance and informational services that they supply for the maintenance of the biological production system that supports human societies.

In essence, unconverted lands contain the output from a four-and-a-half-billion-year-old evolutionary process, which is valuable for the information and insurance contained in the diversity that that process has generated. Evolution generates insurance as a core component of that process; a diversity of life forms has been generated partially in response to an accumulated history of environmental shifts and shocks. Evolution generates information because this range of life forms has co-evolved in the context of biological communities, and the nature of these interactions constitutes a living glossary of the possible forms that biological activity may take. Therefore, diverse resource systems, as the endpoint of this four-and-a-half-billion-year process, contain irreplaceable information and insurance services.

On the other hand, lands that have been subjected to the forces of human specialisation bear little relation to the biosphere that evolution originally placed there. Specialisation involves a clearing of the natural slate, a homogenisation of the environment, in order to allow for the use of the uniform varieties and their ancillary tools. Little of the information and insurance generated by evolution remains.

Therefore, it is the conservation of the output from the evolutionary process that should be the core concern of biodiversity conservation. These evolutionary outputs may be categorised as information value and insurance values, respectively. The general manner in which these evolutionary outputs serve as inputs into the modern biological production system is sketched out below.

2.4.1 Information Value of Diverse Resource Stocks

The mere existence of greater diversity (irrespective of the specific components of that diversity) also has value. One of the most important services rendered by diversity is information. In fact, in Shannon's meaning of the term, the mere presence of variation constitutes information, and uniformity is its absence.

It is also clear that much of the information to be found in biological resources will be particularly useful. This is because the process of evolution guarantees that most biological resources will contain biologically active ingredients (Mabberley, 1992). These types of chemicals necessarily result in the context of interaction within a biological environment. Hence, most plants, insects and animals contain chemicals that act upon the higher organisms that interact with or prey upon them.

The activity of these chemicals can be very constructive in some instances, if properly applied. The knowledge that biological resources have evolved in this way is equivalent to knowing the location of a massive unorganised library on the subject of 'active ingredients'. This knowledge does not identify where to look for a specific ingredient with a specific action, but it does narrow the scope of the search quite substantially away from complete randomness in chemical combinations.

The search for active chemical compounds is an ongoing and well-financed undertaking in the pharmaceutical industry. In 1988, this industry invested over $1 billion in the USA alone on 'Biological Screening and Pharmacological Testing' (Pharmaceutical Manufacturers Association: Annual Survey Report 1988–1990). Although it is possible to synthesise chemical compounds from information derived from the understanding of the biochemical processes of the human body, there appears to be a periodic resurgence in plant-screening for the acquisition of this information (Findeisen, 1991). Only about 5000 plant species have been thoroughly screened for medicinal effectiveness, and 40 of these are in use in prescription drugs. One study identified 25 per cent of all US prescription drugs as plant-based (Farnsworth and Soejarto, 1985). This represents a substantial amount of real economic value in a $10 billion per annum industry (Principe, 1991).

The screening process can be rendered even less random by means of the use of the information accumulated by the human communities living in contact with diverse resources. These people gather this information simply by interacting within their biological environment. For example, the important drug tubocurarine was developed from the poison known as curare used on poison-arrows by Latin American peoples (Iltis, 1988).

Indigenous peoples' information is much more useful than a single example demonstrates. A study by Farnsworth of 119 commercially useful plant-based drugs identified that 74 per cent of them were in prior use by indigenous communities (Farnsworth, 1988). This is what a biologist would expect, and it provides the impetus behind industrial investments in 'ethnobotany', the research into indigenous peoples' traditional medicines (Balick and Sheldon, forthcoming).

The information emanating from diverse resources is used by industries other than the pharmaceutical. For example, many communities raising traditional non-specialised crops have known of useful traits of these species which were not incorporated into the standard commodities. The most closely-related varieties – known as 'landraces' – have periodically been used for improvements to the standard varieties. The crop-breeding enterprise is in fact a major international industry, spending $330 million on the research and development of crop varieties in 1988 (Hobbelink, 1991). Therefore, plant-screening occurs with regard to more than simply medicinal plants.

In sum, biological diversity contains informational value because it maintains a wider choice set. This generates a 'quasi-option value', i.e. the value of retaining a wider set of choices in the event that the decision-making environment 'shifts' (equivalently, information arrives) to render the retained choices relevant (Conrad, 1980).

It remains to explain why it is that information will necessarily arrive in this particular decision-making process. In decision-making over time, 'information' is defined as the occurrence of non-deterministic change in the decision-making environment. The nature of the biological world assures precisely this result. It is the very essence of a dynamic stochastic system, in which the processes of mutation, selection and dispersal continuously alter the natural 'state of nature'. In regard to small organisms, such as bacteria, viruses and insects, these biological processes can occur very rapidly, literally producing thousands of generations in a single year. The biological process is evolutionary, not deterministic, and to the extent that it can be understood, it is too complex to predict. Therefore, in a biological world, the retention of a wider set of biological resources must necessarily have positive informational value.

2.4.2   Insurance Value of Diverse Resource Stocks

Diversity also represents value on account of the contribution that it makes to the aggregate value of a production portfolio. Biological assets are necessarily productive assets, in the sense that they naturally generate growth with time. The tendency towards specialisation in biological assets generates a global production method that is increasingly at risk, precisely because it necessarily generates a narrowing of the global portfolio.

Diverse resources have a role as the providers of insurance for ongoing biological production. Again, this role exists in the first instance by definition. This is because insurance services flow whenever the range of productive assets is broadened. If each of a large number of productive assets has a stochastic element to its rate of return, then (to the extent that these elements are uncorrelated) the return to the combined package of assets will have a reduced variance relative to the variances of the individual components. This is known as 'the portfolio effect', and it derives from the fact that chance variations in output from different assets will tend to cancel each other out if the portfolio is large enough.

Since all biological resources are productive assets (in the sense that they grow and reproduce), the retention of the greatest possible diversity will maximise this portfolio effect, thereby assuring the least amount of risk in biological production. The conversion process, on the other hand, substitutes specialisation for diversity, and therefore would be expected to increase average productivity while reducing the portfolio effect. Table 2.6 demonstrates that, on a global basis and over the 23-year period studied, variability in annual production has been increasing along with the average.

Table 2.6   Impact of worldwide agricultural specialisation on average productivity and variability in productivity (worldwide cereal production, 1960–70, 1971–83)

Years	Average productivity gain (Av. annual rate over 23 yrs)	Average variability (Coefficient of variation)
	2.7%	
1960–70		0.028
1971–83		0.034

Source: Hazell, P. (1989). 'Changing Patterns of Variability in World Cereal Production', in Anderson, J. and Hazell, P., *Variability in Grain Yields*. World Bank: Washington, DC.

This worldwide phenomenon is much more pronounced in regard to those areas and those crops that have been most affected by the process of conversion to specialisation. For example, maize yields in the US were substantially altered by conversion to uniform seed varieties and agricultural methods in the mid-1950s. In the twenty years prior to this, mean yields were about 57 kg per hectare. However, in the thirty years following this conversion, the average productivity of a hectare in maize production increased by over 100 per cent, to 133 kg per hectare

(1955–85). However, the coefficient of variability itself increased by almost 100 per cent (from 0.06 to 0.105) (Duvick, 1989.)

There are two distinct sources of increased variability resulting from increasing specialisation. The first is the loss of the 'portfolio effect' across a geographical region when homogeneous production methods are adopted. In short, when a given territory is converted to the same sorts of crops and methods, the fortunes of all producers then move together. If conditions are favourable to the chosen method, all do better; if conditions are not favourable, all do worse. Since there is no longer as much geographic 'cancelling out', variability is increased. In terms of the proportion of the variability explained, this is by far the more significant of the two contributors. For example, in India, it was discovered that this factor accounted for at least 90 per cent of all of the increase in yield variability (Hazell, 1984).

This increase in variability is an unavoidable consequence of increased specialisation. Increased correlations between the yields of different sites necessarily result from the replacement of differentiated assets with homogeneous ones. These increased correlations then reduce the 'cancelling out' effects of diversity, and thereby increase variability. The trade-off between biodiversity and this form of variability is an identity.

The second reason for increased variability is the less significant contributor proportionately, but the more serious problem. This variability is inherent in the development of specialised species. The reduction of the diversity of the genetic base of a given crop reduces its own resistance to pests. In essence, the existence of variety within the species serves the same purpose of that variety across productive assets: it provides insurance. With increasing genetic uniformity at the level of the species (with regard to various high-yield varieties in use), there is a loss of a 'portfolio' of potential resistance. The specialised, homogeneous crops are consequently more vulnerable to external shocks: pests, droughts, diseases.

For example, in 1970 a particular form of corn blight struck in the US, decimating the crop. Although there were only a few forms of maize susceptible to this pest, a substantial portion of the US crop was planted in a homogeneous strain of one of these types. As a result, approximately 15 per cent of the US maize crop was lost in that year (WCMC, 1992).

As noted, genetic uniformity contributes only a small proportion of total variability in production; however, it is by far the more serious environmental problem on account of the irreversibilities involved (Arrow and Fisher, 1974). Increased variability due to spatial uniformity can be removed at a single stroke, by the reintroduction of more diverse methods at any point in time. In contrast, increased variability resulting from global reliance upon a small number of specialised species cannot be reversed, unless there is a secure genetic bank to turn to in the event of a failure. The existence of diverse genetic resources provides a 'portfolio effect' across time, rather than space. Any narrowing of this portfolio that we undertake now is unlikely to be able to be undone.

In essence, this intertemporal insurance provides a safety net against the possibility of a fatal flaw in any of the specialised varieties. This manner of insurance can be essential for the safeguarding of a population. Anthropologists have hypothesised that the collapse of the classical Mayan civilisation may have been the result of a maize virus. The potato blight in Ireland in the mid-1840s without doubt resulted in the death of over a million people and threatened the collapse of that society (Table 2.7).

Therefore, biological diversity affords a very significant service in the form of the insurance that it provides. A strategy of specialisation must necessarily entail greater risk-taking. Uniform specialisation at the global level would incorporate irreducible and irretrievable risks into the biological production systems.

2.4.3 The Value of Evolutionary Output

There are very real values emanating from diversity: information and insurance. These values derive from the fact that diversity is itself the output received from the evolutionary process. This range and variety of life forms has resulted from millions of years of exposure to a wide range of possible environmental shifts and shocks, and the existing slate of species has coevolved in a manner that guarantees that biological interactions exist. The former facet of evolutionary output (the 'encapsulated history') indicates the existence of insurance services from diversity. The latter facet (the 'coevolutionary action') indicates the existence of informational services from diversity.

Of course, the maintenance of a diversity of life on earth is of interest to the human species as well. The resilience of the human biological production system will be a function of the stability of the underlying biosphere. In fact, it is in the interest of the human species to enhance that systemic stability. It is nearly certain that life itself will continue on earth long after the collapse of the human support systems; all previous incidents of mass extinctions have had much more impact on the higher trophic animal species than on the plant species. The insurance and informational services provided by a diversity of life forms are, therefore, of more than academic interest for the human species.

Table 2.7   Past crop failures from genetic uniformity

Date	Location	Crop	Cause and result	Source
900	Central America	maize	possible collapse of the classical Mayan civilization as a result of a maize virus	(Rhoades, 1991)
1846	Ireland	potato	potato blight led to famine in which 1 million died and 1.5 million emigrated from their homeland	(Hoyt, 1988)
1800s	Ceylon	coffee	fungus wiped out homogenous coffee plantations	(Rhoades, 1991)
1943	India	rice	brown spot disease destroyed crop, starting the 'Great Bengal Famine'	(Hoyt, 1988)
1953–54	USA	wheat	wheat stem rust to most of hard wheat crop	(Hoyt, 1988)
1960s	USA	wheat	stripe rust in Pacific Northwest	(Oldfield, 1989)
1970	USA	maize	decrease in yield of 15%–$1 billion lost	(NAS, 1972)
1970	Philippines & Indonesia	rice	HYV rice attacked by tungro virus	(Hoyt, 1988)
1972	USSR	wheat	crop badly affected by weather	(Plucknett, 1987)
1974	Indonesia	rice	the brown planthopper carrying the grassy stunt virus destroyed over 3 million tonnes of rice	(Hoyt, 1988)
1984	Florida	citrus	disease destroyed 18 million fruit trees	(Rhoades, 1991)
1940s	USA	US crops	losses to insects have doubled since the 1940s	(Plucknett, 1986)

Source: World Conservation Monitoring Centre (1992) 'Valuing Biodiversity' in World Conservation Monitoring Centre, *Global Biodiversity*. Chapman and Hall: London.

2.4.4   Optimal Biodiversity Policy

Human society depends upon the diversity that evolution has generated. However, the nature of these services (diffusive and intangible) makes it difficult for their value to be translated into incentives for investments in stocks of diverse resources. The absence of these incentives is the global problem of biodiversity.

Information is the classic example of a public good. It is so diffusive and non-segregable in nature that it is impossible to appropriate it through a property-based system. The informational values of diverse resources cannot be tied to a particular diverse biological resource, such as a medicinal plant, because the real value lies in the information it disseminates.

Similarly, insurance services from diversity are values that flow to all who benefit from the biological production system, i.e. all of human society. As with informational value, this value is lost once the first sale of a valuable diverse resource occurs. Although all of human society is insured by the presence of diverse resources, there is no mechanism at present for the channelling of this value to the providers of the insurance.

The conservation of diverse resources is not something that will appear to be to the individual interest of a person or state because the services which flow from diversity insure the entire life process, not simply the individual involved. The unchannelled nature of these services renders their values non-appropriable by their host states, and this non-appropriability makes these resources compare unfavourably with other resources. Paradoxically, the very breadth and generality of the benefits rendered by biodiversity (i.e. the safeguarding of the life system) render it an untenable resource under human management.

The solution to the global biodiversity problem is a policy that halts the global conversion process at that point where the value of another conversion (in terms of increased flows of specialised resources) is outweighed by the costliness of that conversion (in terms of lost flows *and stocks* of diverse resources). A stopping rule that puts this into effect is necessary to halt global conversions at the optimal point (Q^* in Figure 2.2).

Such a stopping rule does not currently exist. This is the reason why there is a need for the international regulation of extinction. A mechanism that channels the informational and insurance values of diverse resources through the hands of the host states is the instrument required. This would translate these 'stock-related' values (ignored by host states) into 'flow-linked' values (that command their attention). Since terrestrial resources are all sovereign national resources, the decision regarding their conversion must ultimately be made by the host states. In order to influence their determinations, it will be necessary to shape the decision-making framework within which the conversion choice is made. This must be done by demonstrating to host states that there are feasible development paths compatible with the retention of diversity. For these reasons, the international regulation of extinction equates with the channelling of the values of evolutionary output to those still-unconverted states hosting global biodiversity.

2.5   CONCLUSION

The regulation of extinction is in fact the regulation of the global conversion process. Such a pursuit may sound impracticable or at least highly unlikely, but of course it already exists, only it now functions without consideration of the important global benefits of diverse resources.

These benefits exist because they are the basic output from the evolutionary process. The diversity of life-forms on earth has been the result of four-and-a-half billion years of niche resolution and refinement. This mosaic of life allows the environment to shift quite a lot while still maintaining some successful life forms, thereby assuring the continuation of the life forms under a wide range of conditions.

The human species has brought another fundamental force to earth, i.e. the force of conversion. This force is now reshaping the face of the earth in the space of a few millennia, as fully as the evolutionary force has done over billions of years.

The regulation of extinction is the regulation of this socio-economic process. In particular it is the attempt to channel the benefits of the remaining diverse resource stocks to the states that host them in order to influence their conversion decisions. If it is accepted that all human societies should have an equal right to pursue development, then this is the only means available for influencing the conversion decisions of sovereign states on a year-by-year basis into the foreseeable future. It is necessary to institutionalise a mode of compensation in order to persuade the unconverted states that it is possible to pursue development within the context of diversity.

This is the meaning of the effective international regulation of extinction. It will require significant international institution-building in order to foster these alternative development paths. However, in its absence, the predictable pursuit of the same pathway by successive states will ultimately engender a world devoid of diversity. For this reason, it is especially important to think of this as an opportunity to diversify existing institutions to foster alternative modes of development, rather than the subsidisation of states for non-development. The international regulation of extinction equates with the investment in diversity: biological, economic and institutional.

3 The Economics of Extinction Revisited and Revised

3.1 INTRODUCTION

The economics of species extinction was initially modelled by Clark (1973). (See also Clark and Munro, 1978; Clark, Clarke, and Munro, 1979.) His work provided the framework for most of the analysis that has followed. Given that the past twenty-five years have seen the first efforts to create policies to address the problem of species extinctions, the Clark model has assumed preeminent importance, at the national and international levels. The ideas contained within the models of 'open access overexploitation' have provided the basis for most of the policies for remedying species extinction that currently exist.

However, during this period, the perception of the nature of the problem of species extinctions has altered quite a lot, from well-focused concerns about the endangerment of individual well-recognised species to a much broader concern that includes potential losses of millions of virtually unknown breeds. The latter set of problems is usually categorised as 'biodiversity losses'.

This chapter addresses both forms of extinction problems within a common framework. That is, it constructs a framework for addressing the general problem of species extinctions, entailing the nature of changes affecting habitats and species across the whole of the states of the developing world. This is a general problem that is usually discussed in different contexts, with regard to concerns about rainforests, individual species (such as the African elephant) or general genetic diversity.

It is important that the theory of species extinction is overhauled, as it has generated inappropriate policy conclusions. The main problem with the Clark model is that it placed species-extinction analysis solidly within the previously existing literature on fishery economics. This is problematic because there are far fewer human uses competing for oceanic resources than there are for terrestrial ones. The Clark model simply elided this crucial factor from the analysis.

45

As a result, the policies that were derived from the Clark analysis also ignored this factor, resulting in 'endangered species' legislation that was blinkered against the most fundamental forces for extinction. This has resulted in the creation of some domestic and international policies which themselves represent important threats to many species.

Therefore, the objectives of this chapter are as follows: first, to integrate the problems of endangered species and biodiversity losses; second, to bring the analysis of species extinction back 'onshore' by including competing land uses within the extinction model; and, third, to update the policy framework regarding endangered species as a result of the revised model. The chapter will initially revisit the economics of extinction in section 3.2 by giving an illustration of the Clark model, and its policy implications, in the context of an example regarding the African elephant. Section 3.3 will then integrate the problems of well-known species extinctions such as the elephant with the processes afflicting hundreds of thousands of lesser-known species – the biodiversity problem. Here, the economics of extinction will be treated within a generalised framework of choice regarding productive biological assets. Then, in section 3.4, this investment-based model is revised to incorporate the forces of competition for the terrestrial habitat within its framework. Section 3.5 discusses the implications of these revisions to the economic model of extinctions, while Section 3.6 illustrates the ideas of the chapter in a case study of the African elephant. The chapter demonstrates that the different types of endangerment (overexploitation, conversions, diversity losses) are in fact different routes to the same result, driven by the same fundamental forces. This indicates that the policies for addressing all forms of endangerment, extinction and diversity losses must be reformed to address this fundamental problem. These reforms will often require policies very different from those that have been created in order to address the proximate (as opposed to the fundamental) cause of endangerment.

3.2   THE ECONOMICS OF EXTINCTION REVISITED

Extinction is usually thought of in the context of a few very noticeable examples of endangered species, such as the African elephant. In fact, the decline of the African elephant during the 1980s is a classic example of the ostensibly good fit between the Clark model and actual species decline, as well as the policy implications that derive from that fit. It will be used to illustrate the convincing manner in which this framework misleads policymakers.

During the past decade it is generally estimated that the population of the African elephant declined by about half, from 1 343 340 in 1979 to 609 000 in 1989 (Douglas-Hamilton, 1989). In addition, the trade in ivory, the principal product derived from this species, also doubled between the early 1970s and the early 1980s–from an average of about 550 tonnes to an average of about 1000 tonnes. (Pearce, 1989). On account of these trends, population modellers predicted the imminent extinction of the species, over a period of about twenty years (Renewable Resources Assessment Group, 1989).

How does the Clark model explain the decline and possible extinction of an individual species, such as the African elephant? It does so in relation to three factors:

(1) open access to the resource;
(2) relative price to cost of harvesting the resource; and
(3) relative growth rate of the resource.

These three factors are sufficient to produce a bioeconomic model showing that the optimal stock of the resource goes to zero in the steady-state, so long as: (1) there is open access to the resource for all harvesters; (2) the price of the resource remains high relative to its cost of harvest; and (3) it is a relatively slow-growing resource. In essence, these conditions create incentives for continued harvesting of the resource, even to an extent incompatible with the capacity of the resource to regenerate itself.

3.2.1   A Demonstration of the Clark Model

A simple model will illustrate how these conditions combine to provide the conditions for extinction. Let x represent the stock of a diverse biological resource, where y is the corresponding flow variable (the periodic harvest). The cost of harvesting the resource is then negatively related to the stock of the resource, while positively related to the flow, $c(y, x)$.

Taking an example from Dasgupta (1982), we may say that the cost function is of the following generalised form.

Costs of Harvest

$$c(y, x) = By^{\alpha}x^{-\gamma} \qquad (3.1)$$

Using the first assumption above, open-access conditions of harvest imply that the harvest will continue until the average cost of the harvest of another unit of the resource is equal to its price, i.e. $P = AC$ (Gordon, 1954).

Open Access Equilibrium

$$P = B y^{\alpha-1} x^{-\gamma} \tag{3.2}$$

We also know that the annual change in stock over time, x, will be equal to the intrinsic growth rate of the resource, $H(x)$, less that year's harvest, y.

Annual Change in Stock

$$\dot{x} = H(x) - y \tag{3.3}$$

Combining (3.2) and (3.3) gives an equation that demonstrates that the annual change in the resource stock is: (1) a positive function of the growth rate of the resource; and (2) a negative function of the 'harvesting effort' (which is positively related to the price–cost ratio of harvests). Bioeconomic equilibrium results where these two forces are in balance:

Bioeconomic Equilibrium

$$\dot{x} = H(x) - (P/B)^{\frac{1}{\alpha-1}} x^{\frac{-\gamma}{\alpha-1}} \tag{3.4}$$

A steady-state solution to (3.4) exists where the stock does not change over time, i.e. $\dot{x} = 0$. This occurs where the flow from the growth function is equated with that from the harvesting function. In Figure 3.1, the intersection of these two curves occurs at the stock that will exist in steady state equilibrium.

The broken line in Figure 3.1 indicates the nature of the possibility of 'economic extinction'. In the case of this harvesting function, there is no intersection between the growth function and the harvest function, at any stock level. This means that the incentives to continue harvesting the resource exist at all stock levels, and the growth of the resource is insufficient to maintain a population in the face of these pressures. The basic principle is: low growth rates combined with high price–cost ratios, within the Clark model, result in forces pointing to the ultimate extinction of the species.

3.2.2 The Policy Implications derived from the Clark Model

The policy implications of Clark's model are straightforward. Extinctions result from low growth rates and high price–cost ratios, in the context of poorly managed access. Since there is assumed to be little to be done about affecting resource growth rates, the implication is then to work

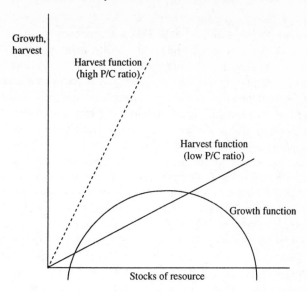

Figure 3.1 Clark bio-economic model of extinction

through the price–cost ratio. That is, the probability of extinction can be
reduced through policies that cause the price–cost ratio to decline. In the
context of Figure 3.1, an endangered species would be found initially in
the situation of a harvest function represented by the broken line. Any act
that would either decrease the price of the resulting products, or increase
the costs of the production process, would shift the harvest function
downwards (to the unbroken line in Figure 3.1) creating the possibility of
a stable bioeconomic equilibrium.

The application of this paradigm leads to some fairly straightforward
policy conclusions. A quick fix for the endangered species within this
framework is the enforced criminalisation of the production process, as
well as a ban on the international trade in its outputs. This policy reform is
believed effective because it is intended to have two important positive
impacts within the context of the model. First, by means of withdrawing
the consumers' support for the species' products, it reduces the demand
and hence the price paid for these products. Second, by means of encouraging domestic enforcement of this criminalization, it increases the
costliness of accessing the resource. Therefore, the policy of enforced
criminalization is intended to have positive impacts working on both sides
of the price–cost ratio. The hoped-for result is that the resulting downward
shift in the harvest function will be sufficient to restore bioeconomic
equilibrium at a stock greater than zero.

This is precisely the policy shift that has occurred as a result of the decline of the African elephant in the 1980s. The international community has acted to impose an international 'ban' on the trade in ivory effective from 1990. Acting within the context of the Convention on International Trade in Endangered Species (CITES), the parties passed a resolution to list the African elephant on Appendix I of that treaty, effectively disallowing all further trade in that species' products (Barbier *et al.*, 1990).

The application of this policy framework is much more general than a single example might indicate however, and the CITES convention demonstrates how pervasive the model's conclusions have become (Lyster, 1985). This treaty is the leading international mechanism for addressing global extinction processes. It is nearly universal in coverage, with more than 120 states party to the treaty. It also applies to a very wide spectrum of species: over 5000 are listed in Appendix I of the treaty.

The policy framework outlined above is also applied at the domestic level in many states. All member states of CITES are required to adopt implementing legislation at the domestic level, and many of these states have extended the prohibitions within that treaty to criminalise internal wildlife use as well. For example, in the US the Endangered Species Act provides that it is illegal to trade in any species, domestic or foreign, that is listed as endangered. Also, throughout many of the states of South America, there are national laws criminalising the trade in all 'wildlife' products. In many of the states of Africa, the trade in wildlife products is also closely restricted or illegal (IUCN, 1985). In fact, the private trade in ivory had been a criminal offence throughout most of Africa for ten or twenty years prior to the elephant's decline.

Therefore, the policies that derive from the 'open-access overexploitation' sort of analysis within the Clark model of extinction have been widely accepted and implemented across the world, both in domestic law and in international law. Across the globe, the regulation of extinction is now handled primarily by attempts to narrow the price-of-product/cost-of-harvest margin in various wildlife and endangered species.

3.3   THE ECONOMICS OF EXTINCTION GENERALISED: BIODIVERSITY LOSSES

Despite the widespread application of this theory of extinction and the worldwide implementation of the policies that it suggests, the natural world is nevertheless heading toward a period of severe mass extinctions. Within the fossil record, there is evidence of about five occasions on which fifty per cent or more of the then-existing species became extinct over a short period of time. Several of the world's most eminent scientists project a mass extinction rivalling these that is set to occur over the next century (Ehrlich and Ehrlich, 1981; Lovejoy, 1980).

Even if the projections of a 'doomsday scenario' are inaccurate, the current rate of extinction is presently far above the 'natural'. Averaged over the 4.5 billion years of biological history, the long-term rate of extinction approximates to 9 species per million per annum. These scientists estimate that the current rate of extinction is about 1000 to 10 000 times this (Wilson, 1988).

Most of these extinctions are not as newsworthy as the elephant's decline, and so they go largely unnoticed. In fact, many if not most of these disappearing species are being destroyed without ever having been classified. This is known as the problem of biodiversity losses i.e. the narrowing of the genetic pool through the losses of many species as well as the loss of much of the heterogeneity within any given species.

This chapter initiates the integration of the framework within which these two problems are being considered. In fact, they are precisely the same general problem, only applied to more and less noticeable species.

3.3.1 The Problem of Endangered Species Generalised

The problems of specific endangered species and general biodiversity losses have their sources in the same fundamental economic problem, viz. the human choice of the set of biological assets upon which society relies. Human societies must select a portfolio of assets from which they then derive a flow of benefits, and one important part of this portfolio is the range of biological resources upon which we depend. These resources are assets (investments which generate flows) simply by virtue of being biological in nature.

Not all assets will be maintained at their existing stock levels. It is sometimes optimal to disinvest in one asset, receive a one-time benefit from that, and invest this receipt in another asset; that is, it is possible to engage in *conversions* between assets. This process will occur until the returns between assets are equilibrated, at the marginal return on capital (*r*) available in that society.

Conversions occur with respect to biological assets as well. Sometimes a biological asset will be converted to a man-made asset (Solow, 1974a). That is, human society will disinvest in the biological resource in order to invest in human-made capital items, such as machinery. Other times a biological asset will be converted to a different form of biological asset, such as the clearing of a forest for a cattle ranch (Swanson, 1990a). The conversion process is at work, even when its workings are not so apparent as the direct conversion of forest land to pasture land. Any asset that yields a non-competitive return will, in the process of equilibriation, experience disinvestment in order to provide resources for investment in the preferred assets, even though the replacement asset may be very different (man-made) or very distant.

It is this process of disinvestment that lies at the base of the decline of all diverse biological assets. Although there are various ways in which this process is manifested (direct conversion, overexploitation, land conversion), the fundamental force determining species decline is the relative rates of investment by the human species. It is the human choice of another asset, over the biological asset, that results in the inevitable decline of species. Extinctions, whether of specific breeds or of general diversity, are the result of their non-inclusion in the human asset portfolio. The remainder of this chapter demonstrates how this is the case.

3.3.2   Species as Productive Assets – The Model Generalised

The generalised model of biological extinction focuses on the investment potential of different varieties of biological resources. It is based on both the Clark model, and the Solow model of asset substitution (Solow, 1974a). It analyses extinction as a process of human choice of the productive assets to retain in the biological portfolio.

The generalised model, building upon that found in the fisheries literature, may be set up as follows (for instance, see Conrad and Clark, 1987). The following variables must be defined, each assumed to be subscripted for time. Also, the variables will all be assumed to apply to a specific diverse resource within a given country. This terminology will be retained throughout the remainder of the chapter, and throughout most of this book.

$p(y)$ the inverse demand curve – a function of aggregate harvest, y

y_i the harvest of the ith agent accessing the resource

$c(x)$ the (constant) average cost of harvest – a function of stock at beginning of the period

$H(x)$ the growth of the resource – a function of the stock of the resource

r the marginal cost of capital within a country

λ the 'shadow price' of a unit of resource stock (equates with resource 'rent' in equilibrium)

At this juncture it is necessary to list a few assumptions in order to define the appropriate societal objective regarding a host state's diverse resource stocks. First, when discussing the problem of global diversity, it is necessary for the discussion to focus on a few states in the developing world, as the vast majority of the world's biological diversity resides in a mere handful of states (McNeely, *et al.*, 1990). Secondly, these states are some of the poorest in the world, and they are therefore justifiably interested in the maximisation of the appropriable values from their diverse resources. Thirdly, the greatest values that are attached to diverse resources generally flow from the developed world, where these forms of resources are relatively more scarce.

This implies that the owner-state with diverse resources is primarily concerned with the capture of the maximum appropriable producer surplus to be derived from its natural assets. This is the societal objective depicted in equation (3.5). From the perspective of this society, species (differing biological life-forms) are differentially valuable as productive assets depending upon the appropriable values that they generate. Given scarce resources to invest, the decision as to which species to maintain in the portfolio of biological assets will depend upon this relative productivity. Therefore, with respect to a given diverse resource (stock x, flow y), the decision as to the stock level to maintain will be made in order to maximise societal benefits from that resource.

Societal Objective

$$\text{Max}_{y} \int_0^{\infty} [p(y)y - c(x)y]\, e^{-rt} \quad dt$$
$$s.t. \; \dot{x} = (x) - y \tag{3.5}$$

This society will solve this problem in order to determine the optimal stock levels for this diverse resource. First, for purposes of later comparisons, the optimal flow (y^*) is shown in equation (3.6), then the optimal stock level (x^*) is depicted in (3.7).

Optimal Flow of Diverse Resource

$$y^*: \; \dot{\lambda} = p\left[\frac{1}{\varepsilon_{D}} + 1\right] - c \tag{3.6}$$

where ε_{D} is the elasticity of demand for the diverse resource.

Optimal Stock of the Diverse Resource

$$x^* : \frac{\dot{\lambda}}{\lambda} - \frac{c'(x)y}{\lambda} + H'(x) = r \tag{3.7}$$

We are concerned here with the steady-state stock levels (i.e. for $\dot{\lambda} = 0, \dot{x} = 0$), because these are indicative of the *societally targeted stock levels* for this particular biological asset. (Of course, this target might take quite a long time to reach, but the entirety of the analysis in this book concerns only the very long run). Equation (3.7) implies that, in the steady state, the resource should be maintained at a stock level that equates the return from that asset with the return from other assets. The return to the stock of a given diverse resource is dependent upon two factors: (1) the relative growth rate (which slows as the stock level reaches *carrying capacity*); and (2) the cost of access – 'search costs' (which decline with increasing stocks). Society will target the stock level (through investments or disinvestments) that generates a competitive return (r).

Figure 3.2 demonstrates this point. An investment-based model indicates that the stock level for each biological resource would be targeted at the level that equilibrates its returns with all other productive assets in the

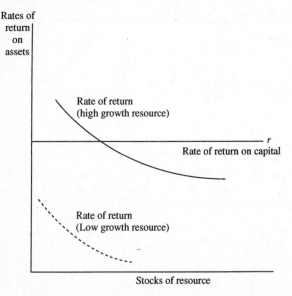

Figure 3.2 Investment-based model of extinction

economy. Since the marginal return on further investments in the stocks of a given biological asset will decline, as the carrying capacity of its habitat is reached, there will always exist an equilibrium stock level for species.

It is clear, within this framework, that the nature of an extinct resource is that of an *non-competitive asset*. The broken line in Figure 3.2 represents a very slow-growing resource which fails to reach the prevailing marginal rate of growth of assets in the economy. In the steady state, there is no stock of this resource that generates a competitive return. Therefore, there is an economic force for the conversion of the entire stock of this asset to other, more productive forms in order to acquire a better return.

Therefore, slow growth relative to other assets in the economy is in and of itself a route to species extinction. Resources, even biological resources, must be competitive as productive assets if there is to be a force for their retention in a world of scarce resources. This point was developed by Clark in the context of his 1973 model. It was further analysed by Clark and co-authors, and several others in a series of articles (Cropper, Lee and Pannu, 1980; Cropper, 1988; Swallow, 1990; Swallow, Parks, and Wear, 1990).

3.3.3   Open-Access Regimes and Investment Incentives

The results derived in the previous section are true even from the perspective of a perfectly well-managed biological resource; non-competitive growth rates are sufficient conditions for asset disinvestment. This point was first discussed in the context of the debate concerning the 'optimal harvest strategy' applicable to the Blue Whale; certain very slow-growing biological resources are subject to pressures for their complete disinvestment (Clark, 1976; Spence, 1975).

Slow growth is not the only route to extinction, however. An alternative route to the creation of a force for disinvestment is the operation of an open-access regime. Open-access regimes can render otherwise viable biological resources largely valueless as economic assets.

The impact of uncontrolled access on a resource can be demonstrated within the generalised model discussed above; this time, however, in the context of a decentralised group of resource harvesters. In this case, each of a large number of competing harvesters (n) attempts to maximise its own profits from individual harvesting activity (y_i). The maximisation problem for each individual harvester is then as follows:

Individual Harvester Maximisation Problem

$$\text{Max}_{y_i} \int_0^\infty [p(y)y_i - c(x)y_i]\, e^{-rt}\ dt$$

$$\text{s.t. } \dot{x} = H(x) - y \tag{3.8}$$

Open-loop Nash equilibrium conjectures will be used to determine how the competing harvesters will use the common resource. Open-loop conjectures imply that harvesters do not take account of *stock effects* in their decision-making. That is, the future impacts of current harvests (increased harvest costs and changed growth rates in the next period) are not considered, because of the small individual impact on the common resource. Nash conjectures imply that there is no strategic interaction; that is, each harvester takes the harvests of others as a given, not influenced by one's own harvest, decision. Open-loop Nash equilibrium conjectures are plausible in the context of an open access regime, precisely because open access implies a large number of competing harvesters.

Under these assumptions, the first-order conditions for an optimal solution to each of the harvester's decision problems are as follows:

Optimal Flow (for Competitive Entry)

$$y_i^* : \lambda = p - c \tag{3.9}$$

Optimal Stock

$$x^* : \frac{\dot{\lambda}}{\lambda} - \frac{c'(x)y}{\lambda n} + H'(x) = r$$

$$(a) \quad \text{or} \quad \lambda = \frac{c'(x)y}{(r - H'(x))n} \tag{3.10}$$

Open access operates to remove all incentives for investment in the diverse resource, and it does so through the process of 'rent dissipation'. That is, under open access, the shadow price of a unit of the resource left unharvested (equivalently, a unit of stock remaining invested in that asset) goes to zero. With a value of zero, there are no incentives for the asset to remain invested in that particular form.

This is because the rent from a natural resource is the wedge between harvest costs and the producer's received price; it is this residual value which is given to the natural resource *itself* by its harvester producer. Of course, other inputs are required in order to produce a natural resource, e.g. the labour and capital required to harvest and market it. If a resource is vastly overproduced (through open-access management), then the only factors earning returns will be the other inputs used in its production. That is, when rents are dissipated, harvesting is occurring until that point where the returns on the other factors are reaching their competitive levels, not leaving a residual for any return on the resource itself. There is then no difference between the marketed price and the combined costs of the *other* inputs into the production process. The rent from the natural resource is being dissipated among the other factors of production.

Equations (3.9) and (3.10) together demonstrate this process of rent dissipation occurring in the context of open access management. From (3.10), as $n \rightarrow \infty$, $\lambda \rightarrow 0$. From (3.9), this implies that $p = c$; that is, the price of the diverse resource is equal to the costs (of the other factors) involved in its production. Therefore, under open-access management regimes, the resource rent value is driven to zero through competitive access.

Open-access management of a resource is inefficient in two respects. First, because the returns from individual investments are distributed across all harvesters of the resource, investors receive an averaged rather than an individualised return, as indicated in equation (3.10). This inefficiently reduces the incentives for individual investments on account of free-riding.

Secondly, under open access, there will be nothing to constrain the number of competing harvesters from increasing boundlessly, in pursuit of any available rents. This rent-seeking behaviour, in the absence of any barriers to entry, must drive the average return from investments to zero.

Therefore, under open-access management, the investment decision devolves to a choice between a return of r (the available marginal on capital) from conversion or a return of zero from investment. Irrespective of the natural capacity of the natural asset for growth, the open-access regime renders it an unworthy investment.

3.3.4   Conclusion – Extinctions as the Result of Conversion

In sum, all biological resources are by their nature productive assets. Assets are kept in society's portfolio on account of the flows they return. Since nature endowed this world with positive stock levels of all biological assets, and human societies now exercise control over all of these stocks, it requires a human decision to not disinvest in order to maintain these. However, investments across assets equilibrate where the returns are roughly equivalent. That is, some assets will experience disinvestment, and others investment, until the returns on assets are equalised. Some biological resources are naturally incapable of meeting the available rate of return, and they are then converted to more productive forms of assets. This force of conversion is the general force that underlies all extinctions.

This process of conversion between assets threatens all biological assets (i.e. species) with depletion in accord with their relative productivities. The general nature of the extinction threat to all species, the more and the less noticeable, is the same. The threat is that a species will be seen as an inferior asset, and thus will be excluded from the human portfolio.

3.4   THE ECONOMICS OF EXTINCTION REVISED

Exclusion from the human portfolio of biological assets is a sufficient condition for biological extinction, but it is not the force that acts directly upon the species to work its decline. Exclusion from the human portfolio is only the most fundamental force at work; it is the 'targeting' of a zero stock level for a given species. This brings into play the forces for conversions; the stocks of the omitted species are then subject to disinvestment.

However, the forces for conversion do not directly work extinctions. There are more consequential actions, as a result of these more fundamental forces, that actually bring about the decline of the species. It is the confusion between the consequential and the fundamental that gave rise to the Clark model (and its policies), and thus it is important to segregate clearly between the two. There are three 'consequential' forces that work species decline. That is, the fundamental decision not to include a specific biological asset within the human portfolio will ultimately affect the stock level of that asset through either:

(1)   stock disinvestments – the removal of the stocks of the asset for sale and allocation of receipts to the acquisition of other, more competitive assets;

(2)   base resource re-allocations – the refusal of an allocation of base resources (land, water, foods) to the asset, and allocation of these base resources to other, more competitive assets; or

(3)   management services re-allocations – the refusal of an allocation of management services to the asset, and allocation of these services to other, more competitive assets.

All of these routes have the same observable effect (i.e. species decline); however, it is more important to show how each has the same fundamental source (i.e. non-inclusion in the portfolio of assets). The remainder of this section demonstrates that the fundamental source of all extinctions is investment-unworthiness; however, the precise nature of the investment required will vary, and this gives rise to several distinct 'avenues to extinction' arising from this common source.

3.4.1 Investment in Resource Stocks

The requirement of asset competitiveness for investment in resource stocks was discussed in the previous section. As indicated in Figure 3.2, if a resource is unable to generate a competitive return, there will be the incentive to remove and market its stocks in order to invest in other assets. This is the scenario of 'optimal extinction' developed by Clark (1973).

Although investment in stocks is necessary in order to avoid extinction, it is not often the case that a species is eradicated by virtue of this manner of stock disinvestment. For example, if there is no demand for the species' products, then there is no incentive for stock-level disinvestments of this type. Therefore, inadequate growth rates do little to explain the nature of the 'biodiversity problem', i.e. the loss of millions of unknown and unclassified species.

Also, there is mounting empirical evidence that the decline of many of the well-known and well-documented species also does not fit this pattern. This is because, although there are clear and substantial markets for these species' products, their values are not being realised by their producers. For example, studies have documented little, if any, rent appropriation by the producers of the products of elephants, rhinos and live birds (Swanson and Barbier, 1992; Swanson, 1991a; Repetto, 1988).

Therefore, the non-competitiveness of the stocks of a resource in terms of growth certainly does generate incentives for direct disinvestment in its stock levels for the acquisition of other assets, but this does not explain a very substantial proportion of the current (and projected) extinctions. In order to explain most species endangerments and general diversity depletion, it is necessary to appeal to other considerations.

3.4.2 Investment in Base Resources for Diverse Resources

A non-competitive biological asset can be disinvested in less direct ways than harvesting and marketing. In particular, all living organisms are dependent upon various biological necessities for their sustenance (food, light, air, water, etc.) Since human societies now have control of these *basic natural resources* (by virtue of their control over land-use decisions), biological assets must compete for an allocation of these resources as well. This implies two fundamental revisions to the basic bioeconomic model.

First, the bioeconomic model of extinction must be revised in order to focus on the fundamental importance of investments in these *base resources* for the survival of terrestrial species. What is required is the jettisoning of an implicit assumption derived from fishery economics, with the effect of bringing the economics of extinction 'on shore'. Specifically, it is important to revise the model in order to provide for the possibility of competing uses for the resources that a species relies upon, and therefore to provide for the necessity of realising a competitive return for the species' use of these resources.

The growth function is the focal point for this revision of the pre-existing model. In general, a logistic form of growth function has been applied for the analysis of population dynamics for most of the past 150 years. This function relates the flow from a stock – $H(x)$ – to the level of the existing stock, x, and to the extent to which that level is less than the available *carrying capacity*, R.

In essence, the logistic growth function implies that if there exists a vacant niche to be filled, the individuals of a species will jointly place more of their aggregate energy into reproduction in order to fill that niche. However, if the species has previously expanded to fill its niche, then most of the energy used by the species will be absorbed in maintenance rather than growth. Therefore, growth rates are linked to the existing stock of a species relative to its available niche. This provides the growth function of the general form, $H(x) = x(R-x)$, with its distinctive logistic shape.

Since the bioeconomic model was developed within the literature of fishery economics, it has always been assumed that the carrying capacity for a given species was a fixed, exogenously given characteristic of the seas it inhabited. And, since humans had few other competing uses for this habitat, the opportunity cost of using it for fish production was assumed to be near zero.

These assumptions are wildly off the mark in regard to terrestrial resources. On land, the amount of habitat available to a given species is endogenous to the process of determining how many individuals of that species will continue to exist, and it is probably the single most important factor determining species viability in the short and medium run. The decision concerning how many base resources to allocate to a species shifts the growth function for that species, determining (in part) the rate of productivity of that species. The nature of that shift is indicated by the two different logistic functions shown in Figure 3.3, each corresponding to the same species, but with different base resource allotments.

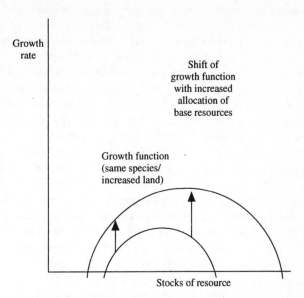

Figure 3.3 Endogenising rates of growth

Therefore, the 'natural' growth rate of a biological resource is affected by the allotment of 'base resources'. However, the provision of these resources habitat is not without a price. The total opportunity cost of these resources equals the flow of benefits expected to be received from the best alternative use of the land, together with the potential returns available from mining the existing stock of natural resources on that land and investing the proceeds elsewhere in the economy. The cost of these implicit investments must be incorporated into the basic model determining the feasibility of investments in diverse resources. (See Swallow, 1990 for related analyses.)

The changes in the extinction model that these alterations imply are as below.

Revised Societal Objective Function

$$\underset{y,\,R}{\text{Max}} \int_0^\infty [p(y)y - c(x)y - r\rho_R R]e^{-rt}\ dt$$

$$s.t.\quad \dot{x} = H(x; R) - y \tag{3.11}$$

where ρ_R is the price of a unit of the base resource R (land)

These alterations indicate that, if this species is to continue to have the use of the resources (R) on which it depends, then it must be able to afford a competitive return (r) on those. The relevant return accords with the marginal rate of return available on investments in this society.

This is now a dynamic model with two decision variables (y and R), and a single state variable (x). One flow variable (y) continues to represent the flow of outputs from the system. However, the new flow variable (R) represents the flow of *inputs* into the system. That is, this variable represents the necessity of continuously *investing* in a system if a flow of outputs is to result. Since (for a stock of any given species to continue to exist) there must be ongoing investment in its resource base, this model captures the importance of this investment.

The implicit assumption in early bioeconomic models was that biological resources were naturally 'free goods' that did not require investment. It was believed that biological product resulted from natural processes of growth deriving from the biosphere's capture of solar energy. This is definitely true in an aggregate sense. The earth does produce a continuous flow of biological product as the result of natural photosynthetic processes.

However, the particular form that this biosphere will take is critically dependent upon relative investment rates, i.e. on the rate at which base resources are allocated to individual species. At the level of the individual species, investment is the crucial decision determining whether a given stock of that species will continue to exist.

A determination of non-competitiveness renders not only investments in stocks of that species inefficient, but also renders inefficient any investments in the base resources required for the sustenance of that species. Without that natural resource base the species will be undercut, resulting in its inevitable decline. Therefore, a determination of asset non-competitiveness is sufficient to determine species decline, but the routes to extinction are more than one.

This alternative route to extinction may also be illustrated in the context of this model. When the additional decision variable (R) is included in this analysis, there is an additional first-order condition that must obtain in the steady state.

Optimal Investment in Resource Base

$$R^*: \quad \frac{\lambda \cdot H_R'}{\rho_R} = r \tag{3.12}$$

This condition states that the a particular species will receive allocations of base resources (approximated by its surrogate – land) only to the extent that the species is able to generate a competitive rate of return from this use. The particular life forms inhabiting a parcel of land are viewed by biologists as a simple outgrowth of that land, using its siting for the appropriation of the sun's energy for regeneration and growth. Equation (3.12) implies that life forms must now compete for their 'place in the sun', subject to human decisions concerning land use. In short, this condition says that the species must compete for the base natural resources which it relies upon; this condition is illustrated in Figure 3.4.

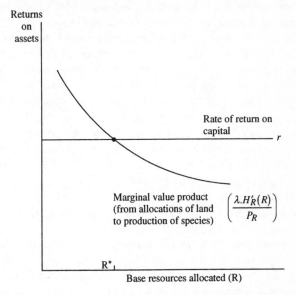

Figure 3.4 Optimal allocation of base resources (land)

Therefore, a non-competitive biological asset is subject to more than one means of removal. It might be removed through an active disinvestment programme (for rent appropriation and asset conversion), but this does not appear empirically to be the most pervasive threat. Most of the extinctions and endangerments occurring today are occurring without notice or knowledge; those that are well-publicised usually occur without the appropriation of many rents. Human-sourced extinctions are working, but not very often through the medium of active, exploitative activities.

In most cases, extinctions are the direct result of passive 'undercutting'. That is, for most threatened life forms, the problem is an unwillingness by humans to invest in the required ancillary resources (such as the terrestrial resources required for biological survival). It is this manner of non-investment that threatens most life forms today.

3.4.3   Investment in Management Services for Diverse Resources

Although the passive 'undercutting' of species through base resource re-allocations probably explains most species endangerments, it does not explain them all. In fact, there is a sub-set of endangerments that fit neither the 'optimal extinction' paradigm nor the 'base resource' paradigm discussed in the preceding two sections. This is the case with regard to the endangerment of those well-recognised species (rhinos, birds, elephants, etc.) whose decline appears to be most closely linked to the 'overexploitation model' discussed above. This is the model developed by Gordon (1954) and Clark (1976) in the context of the fishery, and in which the decline of a species is explained by reference to the existence of an 'open-access' management regime. In the overexploitation model, species decline is shown to be 'caused' by open-access forms of resource management.

In that case, it is important to ask why productive assets potentially worthy of investment would be subjected to such a regime. The answer within this framework is straightforward. From the perspective of the generalised investment model, it is more likely that open-access regimes are caused by decisions not to invest in diverse resources, rather than themselves being a cause of such decisions.

That is, management services are just another form of ancillary resource required for the survival of many forms of life, especially the more noticeable and charismatic species. In fact, these services are just as necessary for their survival as are the base resources of land and water. As demonstrated previously, if there are demand pressures specific to a resource, open-access conditions doom such a resource to disinvestment. The failure to provide management resources for such a resource is equivalent to a decision not to provide an essential requirement for its survival; such a decision is determined by more fundamental considerations.

This point may also be generated within the analytical framework of the investment model of extinction. All that is required is a more complete listing of the essential requirements for the sustenance of a species; these are now seen to be both biological and institutional in nature.

The resources required to sustain a species involve, in the first instance, the natural resource base on which the species depends (R). The next layer of necessary investment is institutional. As indicated above, even with the dedication of substantial areas of lands to certain species, governmental resources are still required for their maintenance and protection. This could be considered in the context of the above model by including another input, M, representing the management services allocated to the resource. These services might be considered a 'stock', in the sense that investments in these services are used for 'institution-building': the creation of a management regime that will yield a flow over the course of many years.

Societal Objective Regarding Diverse Resource

$$\underset{y,R,M}{\text{Max}} \int_0^\infty e^{-rt}\, [S(y;R,M) - c(x)y - r\rho_M M - r\rho_R R]\ \text{d}t$$

$$s.t.\ \ \dot{x} = H(x) - y \tag{3.13}$$

where $S(y; M, R)$ is the flow of social benefits from y *given* the quality of resource management generated by the level of investment in management services (M).

Optimal Investment in Management Institutions

$$M^* : \quad \frac{S_M'}{\rho_M} = r \tag{3.14}$$

Equation (3.14) states that the decision to invest in institution-building with regard to a diverse resource depends upon the perceived benefits that will flow from such an investment. As with all other resources, those

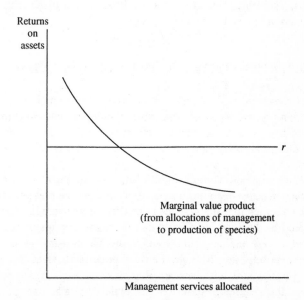

Figure 3.5 Optimal allocation of management services

available for resource management are scarce and competitively allocated. A resource will only receive an allocation of management services, as in Figure 3.5, if it is able to afford a competitive return to these.

This indicates the true nature of the open-access regime. It is not an exogenous 'state of nature', but rather an endogenous 'notion of the state'. That is, open-access regimes are caused by a state-level decision to invest at the lowest possible level for the management of certain resources, i.e. $M = 0$. Under open access, the 'quality' of resource management is at its lowest possible threshold. When this occurs, we are returned to the world of the Clark model (in which 'open access' was the assumed state of the world). In this state of the world, there are no incentives to invest in any way that is associated with the unmanaged species, as the prospects for the return of future benefit flows are severely discounted. Then, all that stands between a resource and its extinction is the net costliness of its harvesting (i.e. its price–cost ratio), as is described in the Clark model.

A positive level of institutional investment (i.e. $M > 0$) is indeed a precondition to all other investment. However, open-access regimes are not base forces for extinction, but merely links in the chain of causation between the base forces and observed extinctions. That is, the same fundamental forces that generate the general incentives for underinvestment (low relative returns) cause underinvestment in management institutions. Therefore, open access is an 'effect' rather than a 'cause' of the fundamental forces driving the extinction process.

3.5   IMPLICATIONS OF THE REVISED MODEL OF EXTINCTIONS

There are two sets of implications to be derived from this revised model of the extinction process. The first concerns the multiple routes to extinction engendered by diverse resource non-competitiveness. The second concerns the different policy implications of the revised model for distinct resources.

In particular, the policy implications of the revised model are, for most species, diametrically opposed to those derived from the previous model. This model implies that the objectives of diverse resource conservation policy should be the creation of basic incentives for investment in diverse species and diverse habitats. This would imply that in many instances the object is to maximise the difference between price and harvest cost, not to minimise it. It would also indicate that most of the current policies concerning the criminalisation of the wildlife trade are probably misconceived.

3.5.1 Alternative Routes to Extinction

In sum, there are three alternative routes to extinction for terrestrial species (as opposed to the single route for marine species). These three routes are:

Stock Disinvestment
As in the bio-economic model applied to marine species, these are resources with high price/cost ratios but low growth. In that case, there are incentives to harvest the entirety of the resource (for its high value) and invest the funds in other assets (for their greater growth rates) (Clark, 1973).

An example of this force in action is the deforestation of the tropical hardwood forests. These trees represent substantial amounts of standing value, but they have very low growth potential. It is economically rational to 'cash in' the hardwoods and invest the returns in other, more productive assets.

Management Resources Diversion
These are resources of 'medium' value but relatively low growth. Since they are slow-growing resources, they make little sense as assets; society has no incentive to invest in their growth capacity. In addition, on account of their relatively low value, they do not justify a commitment of substantial amounts of national resources for the management of the exploitation process. Then, the nation will allow these resources to be depleted through unmanaged exploitation.

Examples of this process include the depletion of many of the large land mammals, such as the African elephant, a case study of which is provided in the next section.

Base Resource Conversion
These are resources that are of little or no known individual value to humans. These biological resources are not overexploited but undercut. They are lost because humans find alternative uses for the lands on which they rely.

An example of this process is the depletion of many types of virtually unknown life forms (e.g. plants and insects) when land is deforested and converted to other forms of use, such as cattle-ranching. This branch of the force for extinction is generally termed the 'biodiversity problem'.

Although there are these three, alternative routes to extinction, they are all consequential forces, not fundamental. As was shown in section 3.4, each of these forces is the consequence of the more fundamental determination that the biological asset is unworthy of investment. It is the particular form in which this decision regarding non-investment first impacts a given resource that determines the applicable route to extinction. However, always at the base is the decision that this particular biological resource is unworthy of human investment.

3.5.2   Rent Appropriation and Investment Incentives

The crucial difference between these models for policy purposes is the impact of an increased price–cost ratio. In the Clark model, as set out previously, the impact of an increased price–cost ratio on wildlife was unambiguous. The effect of increased price–cost ratios was to generate enhanced incentives for harvesting, without a concomitant increase in the resource stock (as the growth function was exogenous to the model). This resulted in an unambiguous increase in the threat to the viability of the species.

In this model, the implications of an increased price–cost ratio could not be more different. In essence, an increase in this ratio implies an increase in rent appropriation by producers. This will enhance the perceived investment-worthiness of the resource.

For example, in the context of equation (3.12), the effect is unambiguous. An increase in rent capture (e.g. resulting in an increase in the equilibrium level of λ in equation 3.12) increases the relative rate of return for this species (i.e. the right-hand side of equation 3.12). More lands may be made available to the species because this species is now better able to 'pay its way'.

This analysis applies equally to the allocation of management services to the diverse resource. So long as social benefits are perceived to be equal to at least the sum of all individual harvests of the species, then an increase in the per unit appropriable value will increase the incentives to invest in this factor as well. In general, enhanced rent appropriation will enhance the incentives to invest in this species.

Rent Appropriation and Investment Incentives

$$\frac{\partial R^*}{\partial \lambda} \succ 0 \qquad\qquad (3.15)$$

Equation (3.15) states that an increase in the per unit appropriated rent increases the level of optimal investment in base resources for the species. This means that, all other things being equal, an increase in rents will increase investment in the species. This would occur by reason of a 'shifting out' of the rate of return curves in Figures 3.2, 3.4 and 3.5. In essence, the increase in rent value has increased the investment-worthiness of the diverse resource.

Of course, this result does depend upon the initial conditions in which the diverse resource is found. If the resource is initially subject to disinvestment by reason of the diversion of management services and it is perceived to be unworthy of investment, then it will already be subject to an 'open-access' regime. Under the circumstances of an open-access regime, a marginally increased rental value will only increase the rate of species decline, while maintaining the inefficient management regime in place.

However, this does not mean that the policy implications of this model are context-dependent, but only that the reforms required may be more drastic in some circumstances than in others. A species that is subject to an open-access regime cannot be saved through a policy of 'rent destruction'; it can only be shifted between the various avenues to extinction. That is, if a species is perceived as unworthy of investment, then a reduction in its marketed value will render it more likely to be lost by reason of base resource conversions and less likely to be lost by reason of management services diversion. This policy cannot save any terrestrial species, but only change the proximate cause of its decline and possibly defer its date of reckoning.

The only policies that can alter the ultimate fate accorded by human society to any particular species is one that addresses the fundamental cause of decline: perceived investment-unworthiness. This implies that diverse resources must be accorded very substantial values, including market values, if they are to receive the investments that they require for their survival. This is the only means by which the fundamental forces driving extinctions and diversity losses might be faced.

3.6   THE DECLINE OF THE AFRICAN ELEPHANT – A CASE STUDY

During the 1980s, the populations of African elephants declined precipitously, according to most estimates. In aggregate, the number of elephants on that continent declined from about 1.3 million to approximately half that, roughly six hundred thousand, in about ten years' time. What have been the causes of this decline, and what are the policies needed to arrest it? This section outlines the nature of the African elephant's decline from the perspective of each of the competing theories above.

Table 3.1   Estimates of African elephant populations
(selected countries)

	1979	1989
Cameroon	16 200	22 000
CAR	63 000	23 000
Chad	15 000	2 100
Congo	10 800	42 000
Equatorial Guinea	1 300	500
Gabon	13 400	74 000
Zaire	377 000	112 000
Central	**497 400**	**277 000**
Ethiopia	900	8 000
Kenya	65 000	16 000
Rwanda	150	50
Somalia	24 300	2 000
Sudan	134 000	22 000
Tanzania	316 300	61 000
Uganda	6 000	1 600
Eastern	**546 855**	**110 000**
Angola	12 400	18 000
Botswana	20 000	68 000
Malawi	4 500	2 800
Mozambique	54 800	17 000
Namibia	2 700	5 700
South Africa	7 800	7 800
Zambia	150 000	32 000
Zimbabwe	30 000	52 000
Southern	**282 200**	**204 000**
Benin	900	2 100
Burkina Faso	1 700	4 500
Ghana	3 500	2 800
Equatorial Guinea	300	560
Ivory Coast	4 000	3 600
Liberia	900	1 300
Mali	1 000	840
Mauritania	160	100
Niger	1 500	440
Nigeria	2 300	1 300
Senegal	450	140

Table 3.1 – *continued*

	1979	1989
Sierra Leone	300	380
Togo	80	380
Western	**17 090**	**19 000**
Totals	1 343 340	609 000

Source: ITRG, 1989. *The Ivory Trade and the Future of the African Elephant*, Report to the Conference of the Parties to CITES, Lausanne.

3.6.1   An Application of the Clark Model?

At first glance, the recent decline of the African elephant appears to be a good example of the workings of the Clark model of extinction. There is no doubt that the proximate cause of most elephant deaths during this decade was a high-powered weapon, nor that the motivation for the slaughter was the procurement of ivory. During the decade the trade in ivory reached a peak of over 1000 tonnes per annum, after averaging nearer 600 tonnes in the previous decade (ITRG, 1989).

It is also clear that the incentives for open-access overexploitation were in place over this period. During the 1980s, the price of elephant ivory soared, reaching prices of more than $140/kg in Japan, while the cost of harvesting (indicated by the prices paid to poachers) was more in the order of $5–10/kg (Swanson, 1989a).

Finally, the growth rate of the African elephant is relatively slow (RRAG, 1989). It has a life-span of 60 years and reaches fecundity at the age of 13 with five-year inter-birth intervals. Studies indicate that a maximum population growth rate of 6 per cent is all that is feasible.

The congruence of these characteristics spells extinction in the context of the simple Clark model. The high price–cost ratio of harvesting combined with the low growth-rate of the species makes extermination the likely economic outcome. In essence, this model would explain the commencement of the downward spiral in elephant populations as the result of a technology shift, i.e. the widespread availability of high-powered weaponry, causing the cost of harvesting to fall precipitously. With this one-time shift in the harvest cost function, it is possible that the new bioeconomic equilibrium would not be established prior to the extinction of the species. In essence, the harvesting function (on account of the technology shift) has shifted up – in Figure 3.1 – resulting in a lower population in the steady state (and possibly extinction).

3.6.2   A More Fundamental Explanation of Species' Decline

An alternative theory would place a species' decline within the context of the broader forces operating across its range. Although the immediate cause of death of each elephant was some hunter's pursuit of its ivory, the more fundamental causes concern the reasons why these hunters were allowed unmanaged access to the elephant herds, and why the ivory harvest was not being managed better for the benefit of the elephants' 'owners'.

This framework provides an alternative explanation for the decline of the African elephant. In short, the African elephant makes little economic sense as a biological asset for investment purposes. Each elephant requires about 50 hectares of good grazing land for its sustenance (Caughley and Goddard, 1975). Average life expectancy is about 55 years (Hanks, 1972). Therefore, it represents a substantial commitment of resources to provide for a single elephant's livelihood.

The resources required for the sustenance of the millions of these creatures that recently roamed Africa would represent a substantial portion of that continent's land area. In addition, few elephants are stationary within an area of a few hectares; they travel widely in search of food, and crops are at particular risk. For these reasons there are substantial negative externalities experienced by those living in the rural areas of a country that has a significant elephant population. Also, the management of access to the population would not be as inexpensive as with a more sedentary animal.

Combined with its slow growth rate and the absence of significant international markets for its products, the pressures for the removal of a substantial portion of the African elephant population from the lands of Africa must be intense. The species will very likely be replaced by a more specialised biological asset, such as cattle or goats or even grain.

In short, elephants do not demonstrate the sorts of characteristics that make an asset worthy of the substantial investments (of natural and governmental resources) that this species requires for its sustenance. This is the fundamental force underlying its decline. The absence of incentives for investment made the species a candidate for non-management.

It is the case that the prevailing management regime in most African ranges states has been *de facto* 'open access'. Open access, in the context of terrestrial resources, is largely a function of the efforts and expenditures of the putative owners. That is, open access occurs when the *de jure* owners fail to allocate sufficient resources to create barriers to accessing the species, or its habitat. In the case of the African elephant, the *de jure* owners are the governments of the African range states, who without exception claim exclusive title to the elephant as 'wildlife'. However, governmental spending on park and habitat protection has been insufficient to regulate access in all but a couple of states, resulting in *de facto* open-access regimes.

This fact has been demonstrated empirically. Poaching pressure and species decline have been shown to be closely related to the governmental spending levels on park protection. In the case of the heavily-poached rhinoceros populations of sub-Saharan Africa, spending on management was shown to be inversely related to the decline of the species' population in those localities (Leader-Williams and Albon, 1988). The equation fitted by these authors indicated a zero population change level at spending of about $215/sq. km. The information that is available indicates that the spending on park monitoring is in fact much lower than this in most African states.

Open-access regimes are better thought of as implicit determinations to not invest in the particular resources, with the object of converting to others that are perceived to be more productive. That is, it is likely to be the decision to deplete a natural resource that generates the open-access regime, not the other way around. Obviously, in the 1980s, few of the range states perceived the elephant as an asset worthy of the investments necessary to maintain existing stock levels.

Table 3.2   Park management spending by selected African states
(spending levels in $/sq. km)

Botswana	1984	10	Niger	1984	5
Burkina Faso	1986	132	Somalia	1984	50
Cameroon	1986	5	South Africa	1986	4350
CAR	1984	8	Sudan	1986	12
CAR	1986	5	Tanzania	1984	20
Ethiopia	1984	57	Tanzania	1986	18
Ghana	1984	237	Uganda	1984	357
Kenya	1984	188	Zaire	1986	2
Malawi	1984	45	Zambia	1984	11
Malawi	1986	49	Zimbabwe	1984	277
Mozambique	1984	19	Zimbabwe	1986	194
Mozambique	1986	7			

Source: Bell, R. and McShane-Caluzi, E. (eds) (1984). 'Funding and financial Control', in *Conservation and Wildlife Management in Africa*, Peace Corps: Washington, DC; Cumming, D., DuToit, R. and Stuart, S. (1986). *African Elephants and Rhinos*, IUCN: Gland.

Unofficial open-access policies have been a good method for mining the vast numbers of surplus elephants, from the perspective of an aid-sensitised African government. The criminalisation of the offtake of ivory preserves international appearances, while the absence of resources applied to elephant protection allows the mining to continue apace. There is, in addition, the side-benefit of the revenues derived from sales of seized ivory. Virtually the entirety of the trade in ivory during the past decade (ranging between 500 and 1000 tonnes per annum) has derived from poached ivory sales that were 'licensed' after seizure. This arm's-length approach to the industry preserves appearances while fostering the removal of the species from the land.

In short, the elephant's decline has been largely the result of an implicit decision to undertake mining on the part of some of the range states. For example, in the 1980s, four countries alone – Tanzania, Zambia, Zaire and Sudan – are estimated to have lost 750 thousand elephants between them, equal to the overall continental losses (ITRG, 1989). Table 3.2 indicates that these states spent $2, $11, $12, and $18 per square kilometre, respectively, on park-monitoring in the year surveyed. The decline of the African elephant during the 1980s, and the ivory trade it spawned, was a direct result of these official non-investment decisions.

Other African states, on the other hand, chose to invest in their elephant populations, and with quite different results. In fact, populations increased by almost 100 per cent in one southern African state that invested heavily in its elephants, Zimbabwe. The relationship is not exact, but it is apparent that most instances of elephant population declines were predetermined by government refusals to invest meaningfully in the species.

This is also consistent with the nature of the observed threats of large-scale losses to less well-documented biodiversity. Throughout Latin America, for example, much of the force for conversions derives from governmental policies that encourage deforestation, and the substitution of specialised agriculture. In Brazil, and many other Latin American states, property-right regimes are only substituted for open-access regimes on the condition that the land is cleared and used for specialised agriculture (Repetto and Gillis, 1988). Throughout many parts of Latin America, it is 'illegal' to trade in all wildlife products, and yet large-scale trade flows from the continent (Swanson, 1991b). The impact of such unenforced laws is simply to enshrine open access with respect to these resources, so that it is impossible for the local populations to foster them as economic resources.

In summary, even in the case of a species such as the elephant, that is apparently well-suited to the overexploitation framework for extinction, the underlying rationale is actually one of underinvestment. This has resulted in the mining of the species from arms'-length over the past decade. The absence of investment incentives has generated this open-access institutional framework, and it has therefore been the fundamental cause of the decline of this species. The adopted international policy, of criminalising the trade and destroying demand, can only bolster the already-existing incentives for the elimination of significant numbers of African elephants from their lands. The fate of the African elephant will be the same, under this policy, only now it will result from 'undercutting' rather than 'overexploitation'.

3.7 CONCLUSION

Even biological resources are now economic resources, in the sense that human societies have control over the terrestrial biosphere and human decisions on resource allocations determine (directly or indirectly) which life forms will continue to exist.

Within this context biological resources are best viewed as a naturally existing form of productive asset, on account of their innate capacity for growth. The human decision within this framework determines the set of productive assets that make up the asset portfolio that human society relies upon. If it is possible for these resources to be included within that portfolio, then it is also possible for them to be left out. Conversion is best viewed as the decision to alter the make-up of the asset portfolio, by means of disinvesting in some assets and acquiring others. The disinvestment decision is the fundamental force that is determining which species are and are not threatened with extinction today.

Although the fundamental force resulting in extinctions is disinvestment, this may occur via a number of different routes: 'mining' for receipts, conversions of base resources or overexploitation. To date, policies addressing endangerment have been focused upon these proximate causes of extinction, rather than the more fundamental one. A policy of this nature can only shift the threatened species between the various routes to extinctions, not alter its ultimate fate. This is precisely the case for many of the species regulated under laws of the nature of the Convention on International Trade in Endangered Species, and species such as the African elephant.

All extinctions are the result of underinvestment. Even overexploitation is the result of underinvestment. Currently, existing policies of demand destruction and supply criminalization are misconstructed and misguided.

They hinge upon crucially incorrect assumptions regarding causation. All extinction policies – those dealing with endangered species, biodiversity policies or land conversions – must deal with the fundamental rather than the proximate causes of the decline of diverse resources. In order to regulate the extinction process effectively, policies must be based on the idea of fostering incentives to invest in these diverse resources.

4 The Global Conversion Process: The Fundamental Forces Driving the Decline of Biological Diversity

4.1 INTRODUCTION

Many leading natural scientists are predicting an imminent 'mass extinction' of species. Some project losses of 25 to 50 per cent of all life forms over the next one hundred years. This concern has been labelled the problem of 'biodiversity losses'. Although these scientists indicate that there are a variety of proximate causal factors regarding extinctions (overexploitation, habitat losses, exotic species), there has been little, if any, work on the nature of the root forces currently threatening extinctions.

The reason for this is that natural scientists view extinction as a natural process, occurring over 4.5 billion years of evolution, whereas the biodiversity problem is better conceptualised as the result of a social process, occurring for very different fundamental reasons over the past ten thousand years. At the time of the climatic upheavals of the late Pleistocene, there was a real discontinuity in the process. Extinctions since that time must be considered the result of a social, rather than a natural, process.

There has been some consideration of extinction within a social framework. Economists have identified two possible fundamental forces for human-induced extinctions; however, to a large extent, these two forces are in logical conflict with one another. One category of forces might be broadly labelled depletion due to the 'mismanagement' of biological resources, the other might be termed the forces for extinction resulting from the human 'management' of biological resources.

Attracting the most attention recently is the problem of mismanagement, and the resulting overexploitation of biological resources. This analysis is the result of the Clark model developed in the context of the fisheries literature (Clark, 1976). It has been applied more recently in the context of tropical forests and many other land-based habitats (Repetto and Gillis, 1988).

Clark also initiated the other line of reasoning regarding human-induced forces for extinction (Clark, 1973). For certain types of species, optimal 'management' policies might themselves lead to complete depletion. In essence, this force for extinction derives from the fact that certain species are uncompetitive as assets, because their natural growth rates are not competitive with the returns available from other assets in the economy. From this perspective, optimal management of the entire portfolio of productive assets implies that some may be relatively poor investments. In those cases it may be rational to 'mine' the whole of a species, and invest the proceeds in another asset. This is possibly the case for slow-growing species, such as the Blue Whale or the African elephant. These species have annual growth-rates of 5 per cent or less, which may not be competitive with the returns available from other investments, and so optimal management points to absolute depletion (Spence, 1975; Dasgupta and Heal, 1979).

However, even these two forces acting together fail to explain the fundamental nature of the extinction process as it is currently occurring across the face of this planet. At present, natural scientists report that endangerment and extinction is not afflicting a handful of relatively slow-growing species, but is a threat to virtually all but a few. The extinction process, as it is presently occurring, approximates much more closely to one of *convergence* upon a slate of a few dozens of heavily-utilised species, with all of the remaining species being relegated to increasingly reduced patches of unconverted habitat.

In order to explain this 'biodiversity problem', it is necessary to derive the conditions under which human choice would generate a mass extinction of species, rather than the selective endangerment. That is, an understanding of the biodiversity problem requires an explanation of *the human actions that homogenise the biosphere*, not an explanation of human actions that degrade the biosphere.

This chapter attempts to provide an answer to the question concerning the fundamental nature of the forces that might generate such uniformity. Specifically, it explains the particular form that the biodiversity problem takes as the result of a dynamic externality incorporated within the idea of 'agriculture', as it diffuses across the globe. This term describes a process that commenced approximately ten thousand years ago when humans first developed the ideas of agricultural technology in the Near East. These ideas, and the tools that they spawned, were used in the cultivation of the various species selected for that purpose in that region (sheep, goats, cattle, barley, wheat, etc.). Over the years, agricultural technology has become embedded in these first-chosen species, so that adoption of these ideas implies adoption of the same species and the same tools used in their production. When the decision to convert to agriculture has been made, it has implied the adoption of the set of ideas, tools and species around which agriculture was originally developed; 'agriculture' is simply the adoption of these specialised methods and species for harnessing the biosphere.

As these technologies have diffused across the globe, it is the universal adoption of these specific methods of production (as embodied within the domesticated and cultivated species) that has produced the homogeneity we now see. It is the inertia flowing from the initial choice of these species, and the many subsequent years of investments, that determines that these are the 'species of choice' in many other environments for which they are not so well-adapted. As the last lands on earth are confronted with the human conversion decision, there is the threat that the last great refuges for diversity might succumb to this process. Human adoption of uniform methods of biological production across the globe has produced the potential for a mass extinction that we are now witnessing.

Section 4.2 of this chapter discusses how the extinction process is now the result of social choice, distinguishing this process from that of 'natural extinction' and demonstrating that the extinction process is now more the result of human management practices, rather than mismanagement. That is, it shows that the true nature of the biodiversity problem is the homogenisation of the biosphere, not its degradation. Section 4.3 presents three simple models of species exploitation, developing the nature of the conversion process. It first reviews the basic nature of the conversion decision, but then demonstrates that it is the 'specialised' nature of conversion that threatens diversity. Finally, it demonstrates that sunk investments in a given technology can create an inertia for the continuing adoption of that same technology; thus, it is the specialised nature of conversion combined with its inbuilt inertia that threatens a mass extinction.

Section 4.4 then discusses the implications of this new theory of extinction, and relates it to the existing evidence on human niche appropriation; in short, the idea of agriculture was an idea on how to advance human wealth and fitness, and to date many of the gains realised have been allocated to the expansion of the human niche. This final section demonstrates that humans have expanded their own niche at the expense of other species, through a 'partnership' with the specialised species. Human niche expansion has occurred, under agriculture, through the selection of a small set of prey species and massive expansions of their ranges. This has created the homogenisation of the biosphere that we are witnessing, and the threat to all diversity that is so imminent. The development and proliferation of the specialised species is the basic force depleting biological diversity. It is itself driven by the human adoption of the strategy of conversion for the advancement of the fundamental pursuit of resources and fitness. These are the fundamental forces driving the depletion of diversity.

4.2   THE NATURE OF THE PROCESS OF EXTINCTION

There are three distinct phases in the history of the extinction process. The first commenced about four-and-a-half billion years ago, and continued until approximately ten or twenty thousand years ago. This was the era in which the natural process of extinction functioned as a fundamental component of the evolutionary process, with little interference from human choice. Evolution operated by allocating 'base resources' (those required for sustenance) to those life forms that were best adapted to the existing niche; those that were less well-adapted were outcompeted, and rendered extinct. The second phase commenced with a technological shift occurring sometime in the climatic upheavals of the late Pleistocene, when humans adopted much more sophisticated technologies for the appropriation of biological product. From this point onwards, human choice usurped the 'evolutionary role', by determining which life forms were to be allocated base resources. Extinction was becoming a distinctly social, rather than a natural, process. Finally, there is now in process a third phase to the extinction process. This is the phase in which human allocations of resources will result in mass extinctions, rather than isolated events of endangerment and extinction. Many scientists predict that the near future holds the potential for a mass extinction of a scale that has occurred only five or six times in the evolutionary history of the planet.

This section describes the fundamental nature of each of these three phases of extinction. Its purpose is to demonstrate that the projected problem of mass extinctions may be conceptualised as the direct result of the management practices adopted by human societies for the purpose of human advancement. In short, many of the gains by the human species over the past ten thousand years (in terms of resources and fitness) have come at the expense of other species. Development and diversity decline have been, to a large extent, two sides of the same coin.

4.2.1 The Natural Process of Extinction

Extinction is itself a natural process and the ultimate fate of every species (Futuyma, 1986; Cox and Moore, 1985). Studies of the fossil record show that the average longevity of a given species lies in the range of 1 to 10 million years (Raup, 1988). Over the four-and-a-half billion years of evolutionary history, the mosaic of life forms on earth has been completely overhauled many times over.

This process is necessary in order to preserve life on earth. The continuous and contemporaneous processes of speciation and extinction form the two necessary prongs of the evolutionary process. It is this constant reshaping of the biological diversity on earth that allows life to continue in the context of a changing physical environment (Marshall, 1988).

Ecologists conceptualise life on earth as a single body. 'Life' is the conduit through which the energy of the sun flows on this planet. It is constantly shifting and reshaping itself, via the evolutionary process, in order to better adapt to prevailing conditions. The various parts (life forms) come and go in order to maintain the integrity of the whole. Any given life form maintains its place within the system solely by virtue of its current capacity to capture some part of the flow of solar energy.

Some life forms are able to capture solar energy directly, simply by virtue of being allotted a 'place in the sun' on earth. These are known as the 'plants', and the fundamental constraint that they face in their competition for existence is territorial. If a plant life form receives an allocation of land, then this is often the only base resource that it requires for survival.

Plants capture the energy of the sun through the process of photosynthesis, and in so doing generate 'Net Primary Product' (NPP), i.e. the product generated from energy captured directly by life forms on earth (Leith and Whittaker, 1975). It is estimated that the NPP on earth is about 225 billion metric tons of organic matter per annum (Ehrlich, 1988). All non-plant life forms must be sustained from this base resource. It is the most fundamental constraint on all other life forms; in order to exist, all non-plant forms of life must receive an allocation of NPP.

This 'allocation' process was a wholly natural one for billions of years. The competition between plant forms for 'land' allocations and the competition between other life forms for NPP allocations lies at the core of the natural process of evolution. These competitions drove the process of evolution, and the evolutionary process gave us the particular set of allocations that human societies 'received' twenty thousand years ago.

Within the evolutionary process, the particular shapes that life takes derive from the relative advantages of these life forms in appropriating some part of the base resource. A 'species' is the name given to a particular form that life takes for this purpose (Barton, 1988). Ecologists conceptualise energy flows to earth as diffusing over uneven gradients, e.g. uneven latitudes and uneven topographies. Given this uneven distribution, base resources will be similarly concentrated; that is, there will be various peaks and troughs across this gradient. A species is defined as the form of life that has established itself as the appropriator of the energy flow across one of these 'resource peaks', or niches. Therefore, the species and its 'niche' are coincident concepts; a species is the particular form that life takes to fit a particular pattern of the energy throughout on earth (the niche).

There is an inbuilt rate of turnover for each niche. First, niches are in a state of constant competition through the process of genetic recombination, mutation, and dispersion. At any time a new form of life may arise that is better 'adapted' to the existing niche, i.e. there is a better fit to the distribution of the base resource. Then the existing life form may be supplanted.

There is another sense in which this process of adaptation must always remain a dynamic one. This is the result of a continuously changing physical environment. For example, the geography of the planet is dynamic. The tectonic plates of the continents have always moved at an average pace of approximately five to ten centimetres per year. These changes create gradual shifts in continental climates. In addition, the earth's overall climate is also in a perpetual state of change. The mean summer temperature in 'Europe' has cycled about 12 times over a range of 10–15 degrees Centigrade during the past million years (West, 1977). The Yugoslav physicist Milutin Milankovich hypothesised that this cycle resulted from the superimposition of three periodic cycles: a 100 000-year cycle resulting from the eliptical shape of the earth's orbit; a 40 000-year cycle resulting from the earth's tilt on its axis; and a 21 000-year cycle resulting from the earth's wobble of the axis of rotation (Hays, Imbrie and Shackleton, 1976).

These in-built mechanisms for environmental variability have generated a system in which the shape and location of any given niche is dynamic. In essence, a niche may be thought of as a peak in the uneven distribution of the base resource, and a species as an outgrowth from the base resource. However, as the topography of the basic resource is not itself a static concept, the process of species definition must also be dynamic.

It is the static result of this dynamic process that can be seen in the distribution of species prevailing at any given point in time, and 'extinction' is the term applied to the changes which occur between static states.

The existence of genetic mutation maintains a pool of potential competitors omnipresent. With the shifting of the underlying base resource, these invaders may be provided with an opportunity at any time. If the invader is successful, and so genetically dissimilar as not to interbreed, then the previous occupant of the niche is superseded, and its range is restricted. If this same result occurs across the entirety of its range, then the totality of its niche is appropriated, i.e. it becomes 'extinct'.

The prevailing distribution of populations is a snapshot of this dynamic competition. For example, the family of Magnoliaceae, genus *Magnolia*, have at present an extremely disjoint distribution. About 80 species exist, with 26 of these in North (and northern South) America and the remainder in South East Asia. However, the fossil record indicates that this family was extremely dispersed during the Mesozoic (100–200 million years ago), also including within their range all of Laurasia (Europe, Asia) through to Greenland (Dandy, 1981).

It appears that the magnolias have gradually succumbed to interspecific competition, i.e. the pressure of faster-growing, better-adapted species capable of better exploiting their former niches. As the environment altered, these competitors appropriated the magnolias' niche throughout Europe, leaving only the disjoint distribution that we currently see. If the process of displacement had continued, the magnolias may have lost the entirety of their niche.

This is the nature of the natural extinction process, as demonstrated by four billion years of evolutionary history. It is not so much a process resulting in the removal of a species, in the sense of niche abandonment, as a process of species resolution, in the sense of niche refinement. It is a natural process for life forms to change, over time and over space. With changing environments, former inhabitants have lost their 'best-adapted status' to invaders, and been replaced. This is recognised as extinction if it occurs across the entire geographic range of the life form. However, in the natural process, it is more accurately conceptualised as necessary turnover deriving from the better adaptation of 'life' to the current state of the niche.

Obviously, this description of the natural extinction process does not accord well with the prospect of losing half of all life forms in the coming one hundred years, i.e. the problem of biodiversity losses. Adaptation is more of a gradualist process in the aggregate sense, even though it represents millions of starts and stops for individual species. Species resolution is a process that occurs over periods of hundreds of millions of years.

This is not intended to imply that all mass extinctions must necessarily be human-induced; they result from any large-scale shock to the life system. There have been several occasions when the rate of extinction of species far exceeded the rate of speciation. There are at least five occasions indicated in the fossil record during which over 50 per cent of the then-existing animal species were rendered extinct (Raup, 1988). These mass extinctions have always been the result of a sudden and dramatic change in the physical nature of the system, e.g. hypothesised sunspots, asteroids, geothermal activity, etc. The dramatic shift of the physical system places a stress on the life system for immediate adaptations. The result of this stress is the loss of many species without their immediate replacement. In essence, this manner of extinction occurs when the niche has been so severely dislocated by a physical event that the species find themselves without the capability to make a claim on the base resource. The 'peaks' in the base resource have shifted out from underneath them.

The current mass extinction is not being initiated by one of these exogenous physical phenomena creating a shock to the system. There is indeed a tremendous stress being placed upon the life system on earth, but it is arising from within the system. This is unique in evolutionary history, and it is not a part of the natural process of extinction. It is a part of the human societal choice process.

4.2.2   The Human-Induced Extinction Process: The Result of Mis-management or Misappropriation?

The natural process of extinction bears little resemblance to the processes occurring over the past few millennia. To a large extent, there is a discontinuity in evolutionary history, arising sometime in the past ten to twenty thousand years. Since that time, there appears to be another layer to the extinction process. There is a record of an increasing number of extinctions, especially of the large land mammals, associated with human arrival or intensive settlement. The change appears to derive from a technological shift, regarding hunting and cultivation, that occurred sometime during the climatic upheavals of the late Pleistocene. From this point onwards, extinction has been more a process of human choice than a process of gradual adaptation.

There is no evidence of human-induced extinctions prior to the evolution of *Homo sapiens* and the introduction of this species to new continents, about 50 000 years ago. Since that time, the coincidence of human migration with species extinctions has been marked. This occurred in Australia (about 50 000 years ago), in the Americas (about 11 000 years ago), in Madagascar (about 1500 years ago), in New Zealand (about 1200 years ago) and on numerous oceanic islands over the past thousand years (Diamond, 1984).

The causes of the more recent extinctions (i.e. over the past 2000 years) have been categorised under three human-induced causal factors: over-exploitation; habitat destruction; and species introduction (Diamond, 1989). However, for the purpose of human choice analysis, it is best to think of all extinctions as arising from one of two fundamental factors: mismanagement or management.

Both potential explanations of the human-induced extinction process are sourced in technological shifts. Extinctions resulting from 'management' derive from technologies (or strategies) that allow one species (or group of species) to appropriate the niches of others. The essence of a management-based extinction is that it occurs by reason of the human species' appropriation of another species' niche.

'Mismanagement'-based extinctions result from a poor fit between human institutions and biological resources. This issue has been the subject of a significant economics literature, primarily in the area of fisheries economics. Recently this theory has also been applied to the mismanagement of terrestrial resources, such as forests and wildlife (Repetto and Gillis, 1988). For example, overexploitation is often theorised to occur when new technology abruptly reduces the cost of harvesting the resource, if social institutions for managing resource access are unable to evolve rapidly enough to forestall the aggregate impact of the new technology. The inability to adapt in time might then result in the complete removal of a given prey species – a jointly irrational step that equates with human niche contraction.

As mentioned, it is also possible to extinguish another species through the process of niche appropriation. This occurs under conditions of 'mutual competition', i.e. where two or more species share a common food supply or breeding ground. The principle of competitive exclusion states that, where two species depend on a common resource, niche expansion by one implies contraction by the other. It is possible for a mixed equilibrium to evolve under these conditions, but if the only basis for the relationship is competitive, then extinction is the likely outcome for one species or the other (Hardin, 1960).

Niche exclusion and appropriation by humans occurs mainly through indirect and surrogate competition. Of course, niche exclusion can occur directly, through the appropriation of another species' resource base for own use (as with the displacement of many American songbirds with the introduction of the European starling). In the case of humans, however, our niche appropriation occurs more indirectly, through surrogates, by the introduction of one human-preferred species in the place of another (as with the displacement of the large African land mammals with selected crops and domesticated livestock).

Niche appropriation, as practised by humans, has consisted of the selection of a few prey species, and the expansion of *their* niches. The discovery of this strategy (domestication and cultivation) and its implementation constituted a very important part of the technological shift that occurred in the late Pleistocene.

Niche appropriation may be used as a strategy to advance either wealth or fitness. In many cases, humans have used the benefits from the expansion of their surrogates' niches for the purpose of their own niche expansion, but this has not always been the case. It is also possible to capture these gains in terms of the reduced allocation of labour to the production of subsistence (and increased allocations of the released labour to other wealth-generating activities), and thus to capture these gains in terms of wealth rather than fitness.

Therefore, there are two competing explanations hypothesised to be at the base of human-induced extinctions: the efficient use of technology for niche appropriation to advance the objects of wealth and fitness accumulation ('management') or the inefficient management of technology with consequent degradation ('mismanagement'). It is clear that a technological shift occurred ten thousand years ago resulting in a new phase of human-induced extinctions; however, whether these extinctions are sourced in the effective or ineffective management of this technological change is less clear. The remainder of this section examines the evidence.

There is no doubt that human mismanagement has resulted in the overexploitation and extinction of many oceanic species, where there is little human competition for the species' niche. Many of the sea mammals (whales, seals, otters) have been hunted to the point of virtual extinction. Stellar's Sea Cow, a slow and easy target, was indeed hunted to complete extinction. These examples most definitely do provide some evidence for the mismanagement hypothesis. However, there is a very important difference between terrestrial and marine resources; the cost of institutional adaptations are much greater in the context of the latter, because no single governance structure applies. In this context, it is predictable that human institutions would adapt slowly to technological changes, to the detriment of the resources.

For terrestrial species, on the other hand, the available evidence (primarily the archaeological record) cannot distinguish the two competing explanations. This is because both potential explanations, management and mismanagement, have the identical objective consequences. Since species and niche are one and the same, the removal of either has the same consequence, i.e. the extinction of the species. Whether humans are pursuing the species or the niche, the observable consequences are identical. In either case, the species will be depleted, and its products may be used in the process; however, in one case the depletion is itself the human choice, in the other the depletion is a by-product of human niche expansion.

The oft-cited 'best evidence' for human mismanagement of terrestrial resources is the coincidence between the introduction of the human species to the American landmass 11 000 years ago and the loss of many of its large mammals; approximately 73 per cent of North American and 80 per cent of South American land mammals disappeared at about the same time as the arrival of Clovis hunters. It has always been assumed that this coincidence implied mismanagement by reason of overhunting, especially as many of the large mammals' remains were evidently slaughtered by humans.

However, substantial climate changes were occurring at this time (Lewin, 1983). These climatic shifts generated a technological shift among human cultures. In particular, it was during such periods of climatic upheaval that the resource management practices known as modern agriculture were developed. The oldest known fossils of domesticated animals are dated to about 12 000 to 15 000 years ago. Widespread changes of diet have been documented for human populations between eight and ten thousand years ago (Davis, 1982; Diamond, 1989). Cultivated grains have been identified back to 6000 years ago (Wright, 1970). Cultivation and domestication at that time implied large-scale land-use changes, especially the widespread use of fire for clearing forests and brush to create agricultural and pasture lands.

It is apparent that a substantial number of human-sourced extinctions occurred throughout the Northern hemisphere ten thousand years ago; e.g. much of the large mammalian fauna was depleted during this period. Despite much speculation, it is not possible for the archaeological record, by itself, to segregate cause and effect in terms of these extinctions. Archaeological evidence cannot say whether these extinctions were the direct result of human mismanagement of the species, or a side-effect of human appropriation of its niche.

However, when the impact of that technological shift on the human species is considered, there is no longer any doubt as to its character. It was at this time (about ten thousand years ago) that the human species' population began to record unprecedented growth. The development of human technologies (cultivation and domestication) in the Neolithic enormously expanded the human niche, from the capacity to support perhaps ten million individuals to a capacity of hundreds of millions in a relatively short time period (Boulding, 1981). This population expansion commencing at this time is a clear indicator that the technological shift of the late Pleistocene was being capably managed and the benefits used for the expansion of the human niche, and the appropriation of many others.

The fact that this population level has been sustained over these ten millennia indicates that this technological shift has constituted a long-term appropriation of other species' niches. The continuing growth of the human population (expected to stabilise only sometime in the next century) is indicative of the diffusion of these ideas of niche appropriation across all of the human societies on earth. In short, human population growth is the best indicator of niche appropriation by this species, and the history of the past ten thousand years indicates that this era has been one of massive niche appropriation by the human species.

In more recent times, when the historical record is more complete, the source of causation is less difficult to discern. Over the past few centuries, the introduction of alien species has become one of the most significant causal factors in species extinctions. For example, over the past four hundred years, there have been approximately thirty species of frog and lizard extinctions, 22 of which (i.e. 73 per cent) have been attributed to the introduction of alien species (Honnegger, 1981). This is a pattern of extinction that is common across many species over much of the last few hundred years; the introduction of aliens is listed as the single highest proximate cause of documented extinctions. The relationship between human management, i.e. methods of production, and recent species extinctions is indicated by the fact that most documented plant and animal extinctions are related to the introduction of only seven alien species: goats, rabbits, pigs, cats and three species of rat (Atkinson, 1989). It has been the human-chosen domesticated species and their attendant 'parasitic species' that have been the source of the majority of modern documented extinction cases.

This effect may also be demonstrated by reference to the study of island habitats. Studies of extinctions on islands are particularly important because they represent a microcosm of the forces occurring globally. Just as most species will have nowhere else to go when human technology diffuses to the final corners of the earth, island species have few options when humans arrive and develop their habitats. The importance of a refuge from human development for species survival is indicated by the fact that fully 75 per cent of all documented extinctions have occurred in island habitats (WCMC, 1992).

Two results from island studies are of particular importance. First, from the fossil record it is apparent that the species diversity of birds on virtually all oceanic islands was reduced by thirty to fifty per cent within the period of initial human occupancy. Since about one-third of all bird species are endemic to islands, this represents a loss of about one-quarter of the world's bird species over the past few thousand years (Olson, 1989). Therefore, the impact of humans over the past thousand years has already worked one mass extinction. This confirms the plausibility of large-scale extinctions as the same forces diffuse as completely on a global basis.

Secondly, it is important to understand the nature of the base forces that have been working this mass extinction. How does the diffusion of human technology impact on a species when abruptly introduced, and how do they react within their threatened habitat? The best evidence shows that species become increasingly concentrated within the least-threatened final refuges.

> Previously it was thought that high islands had greater species diversity because of their montane rain forests. An important observation to emerge from recent studies, however, is that drier, more level lowland habitats, the ones most susceptible to burning and clearing for agriculture, had greater species diversity than steep areas of high, wet forests. On islands, most species that persist in wet montane forests today do so not because this is their preferred habitat, but because it is the only habitat left that has not been too severely modified by man. (Quoted from Olson, 1989).

The diffusion of the technological change associated with human societies might operate globally as it does on these islands. That is, as human technology diffuses to these island systems in the form of land conversions and species introductions, these alterations in turn result in the restriction of the ranges of resident species. If the base resources are all suitable for human appropriation, then the resident species may find themselves without a niche, i.e. extinct. If some part of the base is less suitable for human modification, e.g. the montane rain forests, then the island species become increasingly concentrated in these unaltered habitats, as 'refuges' (Lynch, 1988).

It is probably the case that this same process has been occurring on a global level, over a period of ten thousand years. The global distribution of species is now as much the result of human as of natural processes. Human modification of the environment has restricted the range of many species, forcing the majority into a small number of refugia.

The nature of the fundamental force working during this period is clear, i.e. the transfer of niches to human-selected species in order to expand the claim of the human species on the base resource. Beliefs that mismanagement or human population growth causes extinctions confuse cause with effect. The population of a species is only a measure of its niche; the expansion of the human population from ten million to ten billion is the clearest possible evidence of human niche appropriation. Similarly, overexploitation is often an effect of human niche appropriation; the mining of a species whose niche is gone. Both of these phenomena occur as a consequence of niche appropriation.

Over the past ten thousand years, the course of evolutionary history has taken an abrupt turn. Extinction is no longer a force for better adaptation of all life forms, it is a result of human choices. The choice derives from the technological ability of humans to now appropriate wide ranges of other species' niches. This is the evolutionary nature of cultivation and domestication. Humans discovered how to choose particular prey species on which to build the human niche. Once this was accomplished, the route to human advancement (i.e. human niche expansion) was the expansion of the range of the selected species. It is human cooperation with these species, for the expansion of their ranges, that has resulted in the restriction and extinction of so many others.

In essence, within the natural evolutionary process, the allocation of a portion of 'base resource' (land or NPP) was determined in a natural competition between various life forms. In the past ten thousand years, this allocation decision has been usurped by the human species. Recognising that it was possible to select a species to which to allocate the base resource, and then use that species, the human species had realised a new form of niche appropriation. From that point on, the competition for base resource allocations was a social process.

4.2.3   The Nature of the Current Threat to Biological Diversity

The early, human-induced extinctions have often been those species which are most closely competing with humans, e.g. the large land mammals. The ranges of many of these species are usually restricted by the introduction of humans, and extinction sometimes results. Large land mammals are the most threatened initially, precisely because they compete most closely with humans in terms of range and resource requirements. Other species, with less demanding requirements, have had their ranges vastly circumscribed, or even shifted. As on the oceanic islands, many species have taken refuge in some of the last remaining unaltered habitats. The third phase of the extinction process, i.e. the 'biodiversity problem', is the result of the workings of the human niche appropriation process toward these refugia. As this technological change diffuses to the final corners of the earth, there must necessarily be an increasing rate of extinction.

There is an empirically derived relationship, used by biogeographers, which relates the area of the available unmodified habitat to the number of species that it contains (Williamson, 1988). Studies of 'islands' of natural habitat, whether situated in oceans or civilisation, indicate that the number of species doubles with a tenfold increase in the area of the island. Conversely, a reduction in the size of the natural habitat by 90 per cent is likely to result in a halving of the number of species that it will contain (MacArthur and Wilson, 1967).

Again, the study of islands is instructive for looking at the global impact of conversions. This study, biogeography, has discerned a geometric relationship between conversions and extinctions. That is, the rate of species loss is geometrically increasing with the actual amount of the total base resource appropriated by the human species. As the technology for conversions reaches the final refugia, there will be much greater losses of species per square hectare converted than occurred with the first, conversions. This is what has been occurring on the global level, paralleling the island extinctions from which this relationship was derived.

Land-use alterations have been working across the globe for the past few millennia. Humans have been modifying lands for ten thousand years; however, the pace and diffusion of these alterations have been quickening in the past few hundred years. Estimates of aggregate natural habitat losses over the past two centuries range from 25 to 50 per cent (Myers, 1979; IIED, 1989). Since the commencement of the documentation of land-use changes (in the past thirty years), the pattern of current land conversions is clear. Two hundred million hectares of forest and 11 million hectares of grasslands were converted to specialised agriculture between 1960 and 1980 alone, all of it in the developing countries (see Table 4.1) (Holdgate *et al.*, 1982).

Table 4.1   Rates of conversion of natural habitat to agriculture

	1960 (million ha.)	1980 (million ha.)	Percentage change
Developing			
Sub-S. Africa	161	222	37.8
Latin America	104	142	36.5
South Asia	153	210	37.2
S. E. Asia	40	55	37.5
Developed			
North America	205	203	0.1
Europe	151	137	10.0
USSR	225	233	2.0

Source: Repetto, R. and Gillis, M., (eds), 1988. *Public Policies and the Misuse of Forest Resources*, Cambridge University Press: Cambridge.

In the developed countries of Europe and the US, there is virtually no human niche appropriation still taking place. This is because the process of conversion has been completed. The proportion of Europe which is 'unmodified habitat' (of at least 400 sq. km in area) is now certifiably zero. In the US, a mere 500 years after the introduction of the first Europeans, the proportion of natural habitat of this dimension is down to 5 per cent (World Resources Institute, 1990). The conversion process has worked its way through most of the northern hemisphere, and it is now proceeding in the same manner across the south. At the 'frontier' of this technological diffusion, the rate of conversions remains high.

Table 4.2   Recent rates of conversion to specialised agriculture
(ten-year rate – to 1987)

Conversions to cropland (%)			*Conversion to pastureland (%)*		
1.	Paraguay	71.2	1.	Ecuador	61.5
2.	Niger	32.0	2.	C. Rica	34.1
3.	Mongolia	31.9	3.	Thailand	32.1
4.	Brazil	22.7	4.	Philippines	26.2
5.	Ivory Coast	22.4	5.	Paraguay	26.0
6.	Uganda	21.4	6.	Vietnam	14.0
7.	Guyana	21.3	7.	Nicaragua	11.8
8.	Burkina	19.4			
9.	Rwanda	18.6			
10.	Thailand	17.1			

Source: World Resources Institute and International Institute for Environment and Development (1990).

As the theory of island biogeography, and the studies cited regarding oceanic islands, would predict, this steady contraction of the available unaltered habitat has had the effect of concentrating the remaining species in the remaining unconverted habitat. It has been estimated that about half of the world's species are now contained in the remaining tropical forests. This is in part the result of the workings of four billion years of the natural process of extinction, but it is also the result of the impact of ten thousand years of human-induced extinctions.

Much of the world's remaining diverse life forms exist in one of these last remaining patches of substantially unaltered habitat. The so-called 'megadiversity states', as identified by the World Wide Fund for Nature, are: Mexico, Colombia, Brazil, Zaire, Madagascar and Indonesia; for example, four of these states alone contain approximately 75 per cent of all primate species. In sum, it is estimated that 50 to 80 per cent of the world's biological diversity is to be found in six to 12 tropical countries, including those mentioned above (Mittermeier, 1988).

All of the projections of mass extinctions are based on extrapolations of human land conversion trends into the final refugia of species diversity, i.e. the tropical rainforests. A fairly conservative estimate would seem to be that the diffusion of these technologies into these final refugia is causing current rates of extinction to be about 1000 to 10 000 times the historical rate of extinction (Wilson, 1988). A range of estimates by some of the world's leading scientists projects a possible species loss of between 15 and 50 per cent of the world's total over the next century. (See Table 4.3.)

Table 4.3   Estimated Rates and Projections of Extinctions

Rate %	Projection %	Basis	Source
8	33–50	forest area loss	Lovejoy (1980)
5	50	forest area loss	Ehrlich and Ehrlich (1981)
–	33	forest area loss	Simberloff (1986)
9	25	forest area loss	Raven (1988)
5	15	forest area loss	Reid & Miller (1989)

Note: The rates are given as percentage losses of total number of global species per decade. The projections are based upon the extrapolation of this trend at then-current rates through to the total conversion of the examined forested area.

There is no ambiguity with regard to the source of these projected mass extinctions. The basis for these estimates is invariably the rates of conversion of the tropical forest habitats upon which many of the world's species now depend. The problem of biodiversity, as defined by the world's leading natural scientists, is precisely the problem of the continuing diffusion of human technology to the last refugia on earth. Humans continue to modify the environment in order to replace the pre-existing life forms with those which they have selected. This process commenced ten thousand years ago, and its completion threatens a final human-induced mass extinction. The first acts of niche appropriation had grave effects on those species with wide range requirements; other species, however, merely had their ranges reduced or relocated. Each successive act of appropriation has a greater cost in terms of niche appropriation, and thus increasingly grave impacts on individual species. However, the current problem of biodiversity, as a human-induced mass extinction, is one of the appropriation of the last refuges of the unchosen species. As the diffusion of technology converges upon these sites, the number of species on earth will similarly converge upon the number selected for use by the human species.

4.3   THE SOURCES OF HUMAN-INDUCED EXTINCTION: CONVERSION, AGRICULTURE AND SPECIALISED SPECIES

This section addresses the nature of the social forces that now drive extinctions. This process (of human-induced extinction) is best modelled as an 'asset choice problem'. That is, a species is best conceptualised as a specific type of natural capital, with the capacity to generate its own menu of outputs (goods and services) that is valued by humans. However, the species itself is not the fundamental input; it is merely the conduit through which the fundamental inputs flow. The fundamental inputs for biological production are the base resources that sustain life: land (a place in the sun) for plant life forms, and net primary product for other life forms.

It is the failure to include the competitive nature of the contest for the 'base resources' (land and primary product) within the economic model that has confused the previous analysis of extinction. Of course, solar energy as it arrives here on earth is a 'free good', a natural endowment. However, the specific form that this product assumes depends critically on the array of species that are present to channel it.

The allocation of base resources to particular species is now a societal rather than a natural decision. Since humans do not have the capacity to capture solar energy directly, they must choose which species to use in their consumption of photosynthetic product. And, since humans now have the capacity (through surrogates) to control the allocation of the entirety of the base resource, non-selection equates with extinction. Therefore, it is now this human choice, between species for allocations of base resources, that drives the process of extinction. It is the modelling of this competition, and the forces that drive it, that is the subject of this section.

4.3.1   The Force for Conversion

The process of extinction of species, when conceptualised as specific forms of capital assets, is straightforward. As with all capital assets, there can be no output without inputs. It is the human decision to not supply these inputs that induces extinction. There are three necessary inputs into a biological production system for the capture of photosynthetic product by human societies: stocks, base resource flows, and human management.

The first input upon which all species depend for continued existence and production is their own stock levels. There can be no flow from a species without a prior investment in a positive stock level. Humans must make the decision to invest in supplying these stocks by determining, in each period, whether to disinvest in that stock or not.

Although the stock levels of all naturally existing species have commenced at positive levels by definition, the decision to continue to invest in this array of species is now a human decision, simply because humans have been empowered (technologically) to make the choice. That is, once humans have the power to remove a species (as we do now with regard to all but the smallest), then this option becomes a part of the human choice set, and it must be analysed within this framework. Optimal choice theory will then explain how human societies are making these decisions, not necessarily how they should make them.

The outcome of this decision with regard to any species depends upon the relative return between assets. Stocks of any one species may be disinvested in order to supply funds for investment in another asset. This is the 'force for conversion' between assets, and Solow (1974) was the first to note its applicability to natural resources.

The force for conversion applies equally to the renewable and the non-renewable forms of natural capital. It was first applied in the context of the optimal depletion of a mine (Hotelling, 1931). Its applicability to marine species was later developed by Clark and then Spence in the context of the great whales (Clark, 1973; Spence, 1975). The determinative factor, in the biological context, is the capacity of the natural asset (the species) to compete in terms of productivity with the other assets; therefore, the key to species viability is a rate of growth in excess of the applicable interest rate.

There is another necessary input for all biological assets (in addition to initial stocks). This is solar energy, which is captured either directly (in the case of plants) or indirectly in the form of photosynthetic product (by animals). Solar energy is, as described above, a free good. However, the presence of solar energy outside the earth's system is of little use; there is little usefulness to receiving it on the surface of the moon, for example. The *base resource* through which solar energy is channelled to all terrestrial species is the land itself. The land provides the surface area on which the vegetation resides and accepts solar energy. The land surface area thus provides the channel through which all herbivores accept their solar energy, and similarly for all other life forms.

Each parcel of land constitutes the 'base resource' in the earth's system by virtue of its place in the sun. Organisms which have evolved the capacity to use solar energy require this base resource in order to 'accept' the energy. Organisms, including the human species, that have evolved the capacity to use photosynthetic product must have an allocation of base resource upon which to build their subsistence.

The forces determining extinctions in the modern era are those which determine how humans choose between the various instruments for solar energy acquisition. That is, the earth's surface may be considered the base resource simply because it stands in the path of the sun's energy flow, with an inbuilt capacity to use energy. Each of the ten million or so species may be viewed as a different 'solar panel' from the human perspective, i.e. each species generates an individualised group of goods and services derived from this same input.

This competition between biological species for an allocation of the base resource, land, is another facet of the 'force of conversion'. The more productive species, in terms of the flow of human-valued goods and services, will be substituted for the less productive species. It is the force for conversion between biological assets, rather than between natural and manmade.

These two facets of the force for conversion may be demonstrated in a very simple model, as follows. Consider a human society's decision concerning the allocation of an amount of base resource (R) to a given species (with stock x, and flows y). Assume that it is a developing country whose concern is the maximisation of tradeable product from the flow, and hence the objective is profit maximisation. This choice problem may be considered as follows:

Social Choice Regarding Investment in a Species

$$\operatorname*{Max}_{y,R} \prod(y,R) \equiv \int_0^\infty e^{-rt}(p(y)y - wy - r\rho_R R)\, dt \qquad (4.1)$$

subject to $\dot{x} = H(x,R) - y$

where $\prod$ is the value of the flow of goods and services (y) from a stock of a given species (x)

R is the amount of the 'base resource' (land) allocated to this species

r is the marginal rate of return on investment, i.e. interest rate

$p(y)$ is the inverse (worldwide) demand curve for the flow of goods and services (y)

w is the labour cost per unit of output

ρ_R is the unit opportunity cost of the base resource (land)
$H(x, R)$ is the natural growth function of the species

This model of the decision-making process is chosen for its manner in combining the biological and economic worlds. The foundation of the model is the stock constraint, which represents the character of the biological world. This constraint captures the essence of the biological characteristics of biological production systems. It says that the growth function (H) is a 'free good' derived from the flow of solar energy; but that a flow of the individual species (y) requires prior investments of both stocks (x) and base resources (R). Therefore, biological product flows 'naturally', but the specific constituency of that product is a matter of human choice.

Equation (4.1) gives the criterion by which human societies make their choices regarding relative investments in species. This equation states that, in the case of a developing country (where almost all diverse resources reside), the decision-making criterion will be the rate of return received on investments in that species. In short, in parts of the world where investment deficiencies are obvious and abundant, there will be a need for assets to generate a clear and competitive return.

The society chooses to invest in a given species by its choices of the amount of base resource (R) allocated to the species, and by the amount of offtake (y) of the species. High levels of R and low levels of y represent societal decisions to invest in the necessary inputs for this species (here, stocks and base resources). Then, the state's biological production portfolio will include this species within its array of assets. Any particular species will have to compete to be included within this portfolio, and the nature of this competition can be demonstrated in the first-order conditions of the above model.

The decision on whether to invest in the resource or not lies with the forces determining investments in the two necessary inputs: the prior stock and the base resources.

The General Force for Conversion (stock disinvestment) – 'Stock Investment Criterion'

$$x^* : \frac{\dot{\lambda}}{\lambda} + H'_x = r \tag{4.2}$$

This condition is that derived by Clark and Spence in prior work on the economics of extinction. It states that, in the steady state, the growth rate of the biological resource must be competitive with the general rate of return on assets (i.e. $H'_x > r$ at some stock level), else the society would prefer to disinvest the entire stock of the species and invest the return in other assets.

Another requisite input for inclusion of a given species within the biological production system is the base resource, land (R). Again, the species must compete for allocations of this resource. If no base resource is allocated, disinvestment in stocks occurs more indirectly but just as certainly.

Base Resource Allocation Decision

$$R^* : \frac{H'_R \cdot \lambda}{\rho_R} = r \tag{4.3}$$

The optimal investment condition in equation (4.3) provides that this particular species will receive an allocation of land (R) in accordance with its capacity to earn a competitive return on that allocation. If the species is unable to generate an adequate return, then the base resource will be allocated to another biological asset (or possibly converted to a biologically non-productive activity).

In effect, it is as if the species had to pay a 'rent' each period for the use of the land allocated to it in the amount of the amortised value of the land ($r \rho_R$). Since the price of any particular parcel of land will be bid up in accordance with the potential flow of revenues from its use in production (since it is itself a capital asset), this condition places the species in competition with all other uses of the land for its allocation. The species must be seen as capable of producing a stream of revenues ($H'_R \lambda$) whose present value is sufficient to warrant an investment in the base resource. Otherwise, the land will be allocated to another activity.

This is a fundamental force for extinction. Even if the species is able to compete with manmade capital in terms of productivity, it still must compete with all other biological assets for an allocation of the physical resource. It provides an explanation of why the majority of species are often found only in a few 'refuges' of unaltered habitat. These few sites are situated on what were once considered to be 'wastelands', such as mountains or rainforests, where there has been little human competition for their use. For this reason the price of the lands had remained low, virtually zero, and the majority of species have been confined to these sites in pursuit of a 'low-rent' district', where the competitive forces have been relatively slight.

The recent studies cited in section 4.2, concerning the montane forests on oceanic islands, provide an illustration of this. Although biologists initially believed that these forests were naturally more diverse habitats, it is now believed that these lands contain more species because they have been less susceptible to human modification (burning and agriculture) than are the lowlands. Many of the forest species have migrated to these relatively less-competitive habitats from the lowlands.

The pattern of diversity discovered by modern-day biologists is therefore itself a result of the economic forces for substitution. Species have been confined to 'low-rent' habitats, where human competitive pressures have been lowest.

The same pattern has emerged on a global basis. With the reconstruction of the human asset portfolio, and consequential human alteration of much of the land surface over the past ten thousand years, the unchosen species (unable to compete for investment) have had their ranges restricted to the few remaining refuges of unaltered, undeveloped habitat. The current geographical dispersion of species diversity, as charted by modern-day biologists, must have its source as much in economic forces as in natural ones.

The threat of mass extinctions arises as even these last refuges are coming under pressure for development. It is not the mere fact of a state's 'natural' endowment of diversity, but equally where that state stands in the global development process, which determines the potential magnitude of the impact of these final conversions.

4.3.2 The Nature of 'Agriculture'

The force for conversion is not in itself sufficient to explain the forecasts of mass extinctions. Conversion merely implies that each parcel of land will have its array of biological capital altered in order to maximise its rate of productivity. This means that it is likely that many land parcels will generate a different bundle of biological goods and services, after human modification, but it does not provide an explanation for the reason why many species would not be able to compete for even a single parcel. That is, conversion might imply a redistribution of the earth's species, but not necessarily a narrowing in the total diversity (e.g. the total number of species).

In fact, conversion alone would probably not even redistribute species. By definition, species are best-adapted to their natural environments, and most productive in their own niche (Eltringham, 1984). Therefore, natural productivities (in terms of both growth and human value functions) probably work in favour of maintaining the original distribution of species within the biological production system.

The nature of the additional force working for extinction is made clear by the fact that the entire roster of species is not being considered for use on any given parcel of land. It is more likely that the choice is from among only a handful of 'commercialised' species: domesticated crops and livestock.

The roster of species used to appropriate photosynthetic product for humans has converged to this very small, select group of plants and animals.

This convergence cannot be explained entirely by reason of consumer tastes. There is clear evidence that local tastes favour the local products. Again, this is the expected result of the natural process of co-evolution. That is, it would be expected that the community of predators and prey would co-evolve, so that a given population of the human species would evolve preferences built around their local prey species (Swanson, 1990b).

It is more likely that this convergence is technologically driven. It is the purpose of this section to discuss the nature of the idea of 'agriculture', as it has developed and diffused across the world, and to demonstrate how this idea could generate the sort of global homogenisation of the biosphere that has occurred and that we are witnessing.

The basic idea underlying 'agriculture' is the introduction of capital goods (i.e. tools) into biological production. This allows for the substitution of manmade capital for labour in the capture of biological product. This can be seen both across countries and across time. For example, over the past one hundred years in the developed world, the proportion of the labour force working in biological production has fallen by half to about 3 to 10 per cent with the introduction of modern capital goods (large machinery, chemicals). A contemporary example: at the existing frontiers of agriculture, e.g. Latin America and Africa, the number of tractors has increased by a factor of ten over the past decade, while remaining constant in the developed world (World Resources Institute, 1990). The technological shift which occurred about ten thousand years ago was, in essence, the human appreciation of the fundamental importance of capital goods in biological production. The arrival of this idea opened up a new set of production possibilities for human societies.

It was at this point in time that individual prey species were transformed into distinct 'methods of production' for human appropriation of NPP. Unlike the human inputs for which they were substituting, capital goods were less flexible and adaptable. On account of this, most tools of agriculture were not all-purpose generalists, but were instead 'species-specific'.

This observation is equally true today. The reliance upon capital goods requires homogeneity in inputs and implies the same in regard to outputs. For example, converting a piece of land to the production of a single crop allows the use of machinery for planting and harvesting, and the use of chemicals which are capable of targeting all competitors ('pests') of the chosen crop. All of these inputs are finely-tuned to a single species, which is the resultant (and only) form of output over the entire area of the production unit.

Therefore, agriculture (the idea of using capital goods in biological production) was focused upon the choice between distinct 'species-specific methods of production': consisting of a species and the ancillary resources required (land) and available (capital goods) for its production. The societal choice with regard to biological production possibilities now reduced to a choice between distinct 'bundles' of joint inputs (i.e. methods of production).

The impact of the introduction of species-specific capital goods within biological production may be illustrated in the context of the simple model outlined above. In this version of the model, the role of capital equipment is for the reduction of the average (unit labour) cost of production. That is, the natural production system remains unaffected, but the addition of capital makes the system more efficient in economic terms, by reducing the need for human labour inputs.

Societal Choice Regarding Investment in a Species (with Species-Specific Capital Goods):

$$\operatorname*{Max}_{y,R,K_y} \prod(y,R,K_y) \equiv \int\limits_0^\infty e^{-rt}[p(y)y - c(R,K_y)y - r\rho_R R - r\rho_{K_y} K_y]dt \qquad (4.4)$$

subject to $\dot{x} = H(x;R) - y$

where: K_y is the number of units of species-specific capital goods

ρ_{K_y} is the price of a unit of such capital services

Equation (4.4) captures the idea that the role of agriculture has been the transformation of the societal choice problem to one that incorporates the options that 'species-specific' capital goods (K_y) introduce into biological production. Societies now invest in different species in accordance with how well the overall 'package' of joint inputs performs.

Note that the impact of capital goods in agriculture, in this model, is only to substitute for labour in order to reduce the costs of production $[c(K_y)]$. The fundamental nature of biological production, as depicted in the stock constraint, remains unaffected by the inclusion of manmade capital. This is because agricultural capital goods are not able to affect the rate of flow of solar energy, or the capacity of a given species to produce photosynthetic product from that energy. These capital goods act only to decrease the (units of labour) costliness of planting, removing competitors and harvesting the desired products.

With the introduction of capital goods into the biological production system, the societal choice regarding optimal levels of investment in a species (as a method of production) will now require consideration of two further criteria, in addition to the criteria established in equations (4.2) and (4.3).

Agriculture – Species-Specific Methods of Production
'Scale'

$$R^* : \frac{H'_R \cdot \lambda}{\rho_R} - \frac{c'_R y}{\rho_R} = r \qquad (4.5)$$

'Compatibility'

$$K_y^* : \frac{-c'_{K_y} y}{\rho_{K_y}} = r \qquad (4.6)$$

Equations (4.5) and (4.6) are best conceptualised in reverse order. The 'compatibility' condition states that specialised capital goods will be used in the production of a particular species to the extent that its unit cost function responds to this manner of investment. The 'scale' criterion determines the land area over which the new capital/species production method must operate in order to reach the optimal scale of production for this particular method.

When combined with equation (4.2) above (the 'stock' criterion), these three equations determine the system of incentives for investment in any given species-specific method of production. That is, this system of incentives will determine the optimal method of production to be allocated to any given species, and simultaneously determine the levels of investment in species stocks and in its ancillary resources (capital goods, base resources). The impact of agriculture is that the decision-making concerning the allocation of resources to individual species (and hence the forces effecting extinctions) will be determined within the more general decision-making framework concerning the optimal methods of production to be applied in human appropriation of NPP.

The nature of this decision problem clearly implies societal preferences for some species over others – for the purpose of capital-intensive biological production. First there is clear differential compatibility of species with capital-intensive methods of production; this indicates why the more sedentary ungulates (such as cattle) will be the species of choice over the more migratory ones (such as wildebeest). A small capital investment in a fence is sufficient to maintain a stable production system in the former case, while it is disastrous in the latter. The 'open range' system is a much more labour-intensive method of production than a fenced system, but some species are simply not compatible (i.e. productive) with capital good inputs.

However, the compatibility criterion cannot in itself generate extinctions, because no species requires capital goods for biological survival. It is the combination of the compatibility criterion with the scale criterion *in the context of a strict terrestrial constraint* that generates the outcome of extinction for species.

That is, it is the superior compatibility of certain species with capital goods, together with the increased allocations of land that these methods of production warrant (relative to other species capabilities) and require (in order to achieve a cost-efficient scale of production) that generates the expansion of these species' ranges. The expansion of the ranges of the capital-compatible (or *specialised*) *species* necessarily implies the constriction of the ranges of the less compatible (or *diverse*) species. It is the diversion to the specialised species of the required resources for survival that endangers the diverse.

For example, a large-scale tractor/combine operation may be compatible with a wheat-based production system; however, it might only be cost-effective once a large enough scale of application is achieved (implying large, simultaneous investments in capital, land and stocks). The cost-efficient scale of production, when the costs of the capital goods become great, implies large simultaneous investments in capital, land and stocks for the selected species. As a result, the consequential expansion of the range of this one species, in a world with a fundamental constraint (i.e. the base resource constraint), implies the exclusion of many other species.

Therefore the nature of agriculture implies that the evolutionary prerogative usurped by human society (i.e. the responsibility for allocating base resources to competing life forms) will be made implicitly within the societal decision-making framework concerning the optimal methods of production for use in human appropriation of NPP.

Thus, at the base of the problem of extinction is the human choice between distinct methods of production for niche appropriation. The technological shift implicit within agriculture was the idea that biological production was most cost-effectively undertaken under methods that 'bundled together' the requisite inputs: species-capital goods-base resources. These joint inputs were specialised one to the other, and the choice of a method of production then involved a choice across such bundles. All of the various species on earth have been rendered potential 'methods of production' for the human race, competing for allocations of the ancillary resources that they require for their survival.

With the introduction of the idea of agriculture, the extinction process has since been driven by decision-making concerning the optimal levels of investment in a species and its ancillary resources. An absence of investment in the resources required by a species, in the context of a strict terrestrial constraint, in itself implies ultimate extinction.

4.3.3 Dynamic Externalities: Specialised Species and Learning by Doing

Agriculture cannot in itself explain the phenomenon of convergence within the biosphere. It indicates the reasons why some species will attract human investments, and thereby outcompete other species in the contest for base resources, but it does not explain the number of species that would result from such a competition. Specifically, it does not provide an explanation for why the biosphere would converge upon only a small subset of a few dozen of the available species.

For convergence to occur, it is necessary to include one further factor of production within our model of biological production. This is the factor known as 'knowledge' or 'learning' within the literature (T) (Arrow, 1962). Knowledge in this context is the factor that accumulates with experience or research with regard to the production of a particular species. It is the information that derives from this experience, and that is useful in guiding the production of the species or any of its ancillary specialised factors of production.

For example, it would be possible to retrace the learning curve with regard to many of the specialised species. The first species-specific tools would probably have been hand tools well-suited to the planting and harvesting of the species, e.g. the hoe and scythe for wheat. With experience would come the realisation that further, better tools were possible; for example, the introduction of animal power and the plough in wheat production. On the innovation of the internal combustion engine, the experience with wheat would indicate how best to apply this invention, in the form of tractors, cultivators and combine harvesters. More recently, with the advent of the synthetic chemical industry, the experience with wheat indicated its most virulent competitors (insect, plant and fungus), and dictated the nature of the capital goods (pesticides) to flow from that industry. In sum, the accumulated experience with a particular species would determine, at a minimum, the direction in which new ideas were developed, and very likely determined the direction in which much fundamental research was undertaken.

Thus, this factor is also species-specific in that the knowledge derives from previous work with a particular species and its particular capital goods. It is a factor 'specialised' to a particular species, in the sense that it is costly to transfer the knowledge gained in working with one species across the boundary into other species. It is also specialised in the sense that the work with a particular species determines, in part, the direction in which research is undertaken; the information most suited to the production of diverse species has not even been pursued.

With this final factor introduced, it is now possible to describe precisely the nature of the methods of production available for human appropriation of NPP. These represent the various instruments available for human consideration as means for sustenance.

Species-Specific Methods of Production
$$y = F(x, K_y, T_y, R) \tag{4.7}$$

where: y ≡ the appropriation of NPP in form of species y
x ≡ stocks of species y
K_y ≡ species-specific capital goods
T_y ≡ species-specific knowledge or experience
R ≡ base resources allocated to species y

Equation (4.7) captures the idea of the nature of the contest between species for survival. They are competing as one of four joint inputs within a method of production available for human use in NPP appropriation. Three of these inputs (x, K_y, T_y) are specialised one to another, and together form a distinct 'method of production'. These methods of production then compete for allocations of the fourth, fungible factor (R): the base resources required for biological survival. This competition has replaced evolutionary competition in determining the survival and extinction of species.

The nature of this final factor generates the forces for the convergence of the biosphere upon a small set of specialised species. The dynamic externality inherent within accumulated knowledge and learning generates the nonconvexity within the system, so that human choice falls again and again upon the same small set of life forms. Specifically, accumulated knowledge in this context is a *non-rival good* in the sense of Romer (1987; 1990a; 1990b). That is, it is of the nature of a 'design or list of instructions' that is distinct from the medium on which it is stored, and thus (as pure information) it may be used simultaneously by arbitrarily many agents without added cost. The accumulated experience in regard to the specialised species is inherent within the capital goods and species as they stand, and is available at no added cost to the marginal user (Romer, 1990b).

This dynamic externality creates the non-convexity within the system. Specifically, it is attributable to two different 'spillover' effects regarding accumulated agricultural knowledge, one that does and one that does not exist. First, it is costly to transfer accumulated knowledge and experience across biological boundaries (spillover between species does not exist). Second, it is not costly to transfer accumulated knowledge and experience across geographical boundaries (spillover between users does exist).

The combination of these two effects would create a clear non-convexity within the global system of choice regarding distinct methods of production for appropriating NPP. Essentially, there would evolve through cumulative causation an inertia around the initial conditions of agricultural technology by virtue of the accumulation of experience around these initial conditions, and the non-costly transfer of this experience to subsequent users. The result would be an example of 'path dependence', or a situation in which history 'matters' (i.e. the initial conditions determine the subsequent path) (David, 1985).

This non-convexity is of two types. First, the idea of agriculture would arise in a given location, by the choice of particular species for development in combination with the newly-introduced capital goods. With cumulated experience with these particular species and their associated tools, a stock of species-specific knowledge is generated. This stock of knowledge is productive in two ways: it contributes to the cost-effective production of the chosen species (given existing technologies) and it also contributes to the cost-effective production of new species-specific capital goods. The cumulation of this stock of capital thus represents the increasing returns to scale that derive from learning by doing (Arrow, 1962). It creates the incentive for systems that have adopted a particular set of production methods to continue down the same path.

Secondly, as the idea of agriculture (i.e. the use of species-specific capital goods) diffuses across the globe, there is a nonconvexity in the choice set afforded to states on the agricultural frontier. This results from the non-rival nature of the existing stock of knowledge associated with the specialised species (Romer, 1990b). In essence, the state on the frontier has a choice between *adopting agriculture* in its species-embedded form (and thus receiving the stock of knowledge inherent in these species and their capital goods at no extra charge), or in *adapting agriculture* to its resident species (and thus facing the requirement of generating its own stock of knowledge at its own expense).

Therefore, the combination of these two types of dynamic externalities creates the force for global convergence upon a particular set of species for human appropriation of NPP. The increasing returns to scale derived from learning by doing keep the early converting states on the path initially selected. The non-rival nature of this knowledge provides the incentives for later converting states to follow the same path. Together, these forces provide the incentives for the human-constructed biosphere to converge upon that set of species within which agriculture was first initiated.

The initial 'chosen' species were selected from a small sub-set of the global number, dependent upon the locations in which this technological shift occurred. Investment was then 'sunk' into these species, as the capital developed has been finely-tuned to their specific characteristics.

Finally, as the idea of agriculture diffused, it did so in embedded form, i.e. the idea of capital-intensive production of species diffused in the form of already-selected plants, animals and equipment. Therefore, the acceptance of the idea involved the conversion of the habitat (and resident species) to the already-specialised ones.

The dynamic externality implicit within the idea of agriculture may be demonstrated within the context of this simple model. The relevant question concerns the choices made by those individuals and societies on the agricultural frontier; that is, what forces will determine the nature of conversion at the global frontier? Consider a single state (s) attempting to decide in which species to invest, financial and base resources, in its transition to agricultural production. The only change to the previous model is the inclusion of the additional factor input, knowledge (T_y), within the decision-making framework of the frontier state.

State's Choice of Methods of Production in Agricultural Transition

$$\underset{y, R, K_y, T_y}{\text{Max}} \prod_s (y, R, K_y, T_y) \equiv$$

$$\int_{t=0}^{\infty} e^{-rt} \left[p(y)y - c(R, K_y, T_y)y - r\rho_R R - r\rho_{K_y} K_y - r\rho_{T_y} T_y \right] dt \qquad (4.8)$$

subject to : $\dot{x} = H(x, R) - y$

where: $T_y \quad \equiv$ *stock of species-specific knowledge or experience*

$\rho_{T_y} \quad \equiv$ *price per unit of species-specific knowledge or experience*

Once the importance of this factor is recognised within the framework of the marginal state, there is an additional investment criterion to be included with those previously derived (4.2, 4.5 and 4.6) in determining the optimal methods of production to be applied. That is, the newly-converting state must consider not only the capital-compatibility of various species in choosing the methods of production to apply, it must also consider the availability of knowledge and experience in the use of the species (or the costliness of developing the same) and the cost-effectiveness of this knowledge. This will be referred to as the condition determining the 'optimal stock of learning' with regard to a species-specific method of production.

Optimal Stock of Learning

$$T_y^* : c_{T_y}' y = r\rho_{T_y} \qquad (4.9)$$

Equation (4.9) states that the optimal level of learning with respect to any species-specific method of production will equate the marginal benefits from additions to the stock with the marginal costliness of these additions. The benefits to learning supplements are straightforward. In this model they act to reduce the (unit labour) cost of production of the appropriable biomass (y), by substituting for the other factors of production.

The non-convexity of the state's choice set derives from the relative costliness of additions to the state's stock of learning. This derives from the alternative methods for producing this knowledge. One means of producing learning is by pre-production research or on-line experience. Each of these alternatives is costly. The former implies postponing the choice into the future while learning is acquired through research; both research and postponement are costly. The latter implies the accumulation of experience *after* the choice of method has been made; this implies commencement with a zero stock level of learning and the incurrence of sunk costs for the selected species.

The alternative for the state is selection from the slate of the already-specialised species, which incorporate the accumulated learning from their usage within their specialised capital goods and within themselves. For these species, the existing stocks of learning are non-rival goods, available at no additional cost to the marginal users of these species. That is, the long history of usage for these species has resulted in informed choices that have been incorporated within the method of production of which these species are part. Selecting these species brings with them this accumulated knowledge. Selecting other than these species implies the necessity of incurring the costliness of producing these stocks, or forgoing their benefits.

Costliness of Alternative Methods for Acquiring Stocks of Learning
Acquisition of learning by either:

(1) Imitation $\rho_{T_{y_i}} = 0$ for $\quad 0 < T_{y_i}^s \leq T_{y_i}^e$; **or**

(2) Research $\rho_{T_{y_i}} > 0$ *since* $T_{y_i}^e = 0$. $\hspace{2cm}$ (4.10)

where: $\quad T_{y_i}^s \quad \equiv \quad$ the stock of species-specific learning acquired by state(s)

$\qquad T_{y_i}^e \quad \equiv \quad$ the existing (global) stock of learning for species y_i (specialised or diverse)

$\qquad \rho_{T_{y_i}} \quad \equiv \quad$ the cost of species-specific learning

Equation (4.10) states the reason for the non-convexity in the converting state's production possibilities set. It arises from the non-rival nature of the stock of global experience with the specialised species. This represents an implicit subsidy to those producers adopting these species-based methods of production.

Since the diffusion of agriculture has taken time (approximately ten thousand years), the accumulated learning regarding the previously selected species has generated a virtually irresistible force for the currently converting states. The stocks of the specialised species (cattle, sheep, pigs, buffaloes) have risen to such levels (1.3 billion, 1.6 billion, 800 million, 150 million, respectively) (World Resources Institute, 1990) over such a long period (ten thousand years) that it would require virtually limitless amounts of resources to replicate these levels of cumulated experience for other species in a shorter amount of time.

The capacity for 'learning-by-doing' forms of dynamic externalities to generate natural monopolies has already been demonstrated in the context of the 'experience-advantaged firm' (Dasgupta and Stiglitz, 1988). In short, if there are sunk costs involved in accumulating learning (i.e. the benefits of learning do not spill over between production units), then the most cost-efficient form for the industry to take is a single production unit. In the private sector, the industry would probably assume an oligopolistic structure (Dasgupta and Stiglitz, 1988, p. 247).

In the context of our model, the 'industry' concerned is the human quest for cost-efficient biological production, and the 'production unit' is a species-specific method of production. The selection of an initial slate of species at the onset of agriculture vested these species with the advantage of experience, and this advantage has acted as a barrier to entry for other species-based methods of production. This is how this initially chosen slate of species became 'natural monopolies' as methods for producing the biological product that human societies rely upon.

Therefore, the idea of agriculture has become embedded in these few species, and it has diffused as a package of specialised species and species-specific capital goods. That is, as the idea underlying this technology has globalised, so have these species. Rather than applying the idea of specialised production to the naturally resident species, the non-convexity within the system has dictated that the lands be 'converted' to the previously-selected package of species and capital.

The result has been a two-sector biological world: one homogeneous in its roster of species and methods of biological production, and the other diverse in the same. As the idea of agriculture has moved across the face of the globe, it has shifted the boundary between the homogeneous and the diverse sectors, continually restricting the diverse sphere to an ever smaller area, until now we are faced with two closely-linked outcomes.

First, the biological production 'menu' has converged to a relative handful of species. Of the thousands of species of plants that are deemed edible and adequate substitutes for human consumption, there are now only 20 that which produce the vast majority of the world's food. In fact, the four big carbohydrate crops (wheat, rice, maize and potatoes) feed more people than the next 26 crops combined (Witt, 1985). There are now fewer than two dozen species which figure prominently in international trade (FAO, 1988). In short, humans have come to rely upon a minute proportion of the world's species for their sustenance.

Secondly, a substantial number of the other, unchosen species are endangered by reason of relative underinvestment. The continuing diffusion of this technology to the final corners of the earth threatens to place the entirety of the globe into a single, homogeneous sector in the coming century; the area of unconverted habitat dwindles each day. The ongoing conversion to a very small number of species leaves the vast number of species without a resource base, as they are undercut by the expansion of the niches of the previously-selected plants and animals.

This is the nature of the human-induced extinction process, as it is presently occurring. This is a by-product of the technological shift that occurred ten thousand years ago, when humans first selected species for niche expansion. The current threat to biodiversity is the logical result of the diffusion of this technology to the ends of the earth.

4.4. THE IMPACT OF SPECIALISED SPECIES – AN EMPIRICAL EXAMPLE

The ideas of agriculture have been applied by the human species for almost ten thousand years. On account of the dynamic externalities within this process, the gains in productivity have come at the expense of losses in variety. As this technological change works its way to the ends of the earth, the aggregate impact on the remaining diversity is steadily increasing.

The driving force behind the use of this strategy is the human drive for increased wealth and fitness. The cumulative impact of this process is reflected in the expansion of the human niche as well as in human resource accumulation. However, there are other strategies that are able to contribute to human wealth accumulation, while only niche appropriation can contribute to human niche expansion. The evidence on human niche expansion is the clearest we have for the measurement of the nature and scale of this technological shift.

This impact is demonstrable in two ways. First, as previously mentioned, a species' population is a measure of niche, and under this definition the human niche has increased by three orders of magnitude since onset of this technological change. Although we have become accustomed to living in a state of disequilibrium, it is necessarily the case that the continuing growth of the human population comes by means of the appropriation of other species' niches.

The relative success of species in the competition for base resources is the other evidence we have on the impact of human niche appropriation. This is reflected in the distribution of Net Primary Product between different life forms. Current estimates indicate that the human species, one of about ten million species, sequesters about 40 per cent of terrestrial NPP to itself. Of course, this level of niche appropriation in and of itself excludes many other life forms. The magnitude of human NPP appropriation is an important threat to global biodiversity.

Table 4.4   Human Appropriation of The Earth's Biological Product

Form of human use	Share of NPP (%)
Direct use	4
Indirect use	26
Losses	10
Total	40

Source: Vitousek, Ehrlich, Ehrlich, and Matson (1986)

It is none the less important to distinguish between the effect of 'pure scale' and other causes active in the depletion of diversity. The human species cannot live without prey species, so human population expansion (pure scale) does not necessarily equate with the depletion of diversity, of the nature currently witnessed. That is, it would be possible for the human species to vastly expand its niche by means of excluding only those species in closest competition (i.e. those at or near the same trophic level). If the nature of the diversity problem were pure scale, then it would be expected that humans would be removing the species at or near their own trophic level (the large land mammals) and concentrating on the more efficient methods of capturing NPP (i.e. the substitution of plants for animal prey). Appropriation methods of this nature might deplete all of the land mammals, and most other predators, but they would not threaten a mass extinction of all forms of diversity (since the vast majority of life forms would be potential prey rather than competing predators).

It is not human population expansion by itself, but the particular form that this expansion has taken (i.e. the homogenisation of human prey) that has been most destructive of diversity. That is, human population expansion is more a consequence than a cause of diversity depletion. The more fundamental factor is the human-sourced homogenisation of the biosphere for the cost-efficient appropriation of biological product. This homogenisation *directly* depletes diversity and *consequentially* enables further expansion of the human population.

In order to make clear the distinction, consider the consequences flowing from the discontinuance of either process: the homogenisation of the biosphere or the expansion of the human population. To date, the two processes have gone hand-in-hand, i.e. biosphere homogenisation has equated with niche appropriation, but this need not necessarily be the case. It is possible to unlink the two processes and thus it is possible to continue one without the other.

First, consider the potential halt of biosphere homogenisation. It remains possible for human populations to continue to expand without the depletion of remaining diversity. All this requires is that agriculture be adapted rather adopted, i.e. the stock of learning for the use of diverse species must be developed rather than adopted in the form of the specialised species. It is more costly, but still feasible, to expand the human niche without necessarily substituting specialised for diverse species.

Second, consider the possible halt of the process of human population expansion. It remains likely even in this event that diversity will continue to be depleted. This is because biosphere homogenisation is a generally cost-effective measure for biological production; it generates wealth for the human community through cost-effective factor substitution, irrespective of its scale. To date, much of this increase in human wealth has been absorbed in terms of population expansion (to the dismay of many human communities), but this does not imply that halting population growth will equate with halting diversity depletion. To achieve this result it is necessary to assume that human societies will halt processes aimed at wealth expansion as well, an unlikely prospect.

The relative importance of the various contributing factors to diversity depletion is indicated by the various categories of current human use of NPP. The impact of the scale of the human population is indicated by the amount of direct use, which is 4 per cent of NPP. While this is in itself substantial, it is not out of line with the complexity and trophic level of the species (Wilson, 1988). Therefore, the 'scale' of the human niche (and hence its population) is not, as indicated by direct NPP consumption, sufficient to be a threat to global biodiversity at present.

A more significant force placing pressure on other species is human 'wastage' of NPP. This results most dramatically when good soil is paved for a road, but it occurs more commonly when a forest or pasture is converted to a common commodity such as corn. The homogenisation of the system of outputs to a single species has the ancillary effect of reducing the total biomass of the system. Species are expert at finding an available niche in which to prosper, but they require variety in the environment in order to distinguish themselves. For example, a tree by itself creates a vertical dimension to an otherwise horizontal environment and can thereby generate innumerable niches. Clearing the natural environment removes a large number of such niches, thus resulting in a net loss in ecosystem product. Therefore, an even greater force for human-induced extinction than human consumption arises from human reliance upon homogeneous outputs in biological production.

However, the most significant force for extinction derives from the amount of ecosystem services which humans use 'indirectly'; this constitutes fully 26 per cent of total NPP. These uses include the parts of the plants that we grow which are not consumed by us, and thus return to the earth unused by a higher organism. It also includes the amount of biological matter which is cleared and burned in the agricultural process. These are uses that derive from the fact that humans require homogeneous inputs in a system that produces homogeneous outputs. Lands are cleared and levelled. Other species are denied access as potential pests are predators. The inputs into the system are restricted as far as possible to those entirely within human control.

Of the 40 per cent of NPP appropriated by the human species, 4 per cent is attributable to consumption and 36 per cent is attributable to methods of production. The idea of agriculture has had its impact by virtue of replacing heterogeneous production systems with homogeneous production systems, and this constitutes 90 per cent of the 'niche' currently sequestered to human use. This is the core of the problem of biodiversity endangerment.

4.5 CONCLUSION

There have been three different phases to the process of extinction. The first phase extended across nearly four-and-a-half billion years, and was a fundamental component of the evolutionary process on earth. The two other phases commenced only ten thousand years ago, and are processes of human-induced extinctions for the expansion of the human niche.

Ten thousand years ago, the idea of domestication occurred to human communities. The kernel of this idea is for the human species to select certain plants and animals, and specialise in the consumption of their products. To a large extent this idea set the human species free from its previous niche, and made them reliant instead on the niches inhabited by the selected species. For the first time in evolutionary history, the fates of different species became *voluntarily* interconnected.

Rather than choosing different species to partner in different habitats, humans have instead chosen to spread a handful of select species across the globe. This is because investments in species-specific capital goods are sunk, and the productivity gains of 'agriculture' derive from combining capital goods with species in biological production. With specialised 'learning-by-doing', subsequent adoptions of this technology are biased towards the originally selected species. The original developers of the technology continue to invest in it on account of the dynamic returns to scale. The newly-converting states adopt it on account of the non-rival nature of the knowledge. For these reasons, the technology has come to be embedded in a handful of plants and animals.

The primary results of this new strategy of inter-species cooperation have been two. First, the human species and its specialised prey have had their niches expanded beyond recognition. The population of the human species has expanded from ten million to ten billion in ten thousand years' time.

The past ten thousand years has also seen the human species work hard for the expansion of the niches of these specialised species. There are now more than one-and-a-third billion cattle on the earth, inhabiting every continent on the earth. The human species has selected just four species of plants (potatoes, rice, wheat and maize) upon which it relies for the majority of its daily diet. These species have also seen their niches expanded to the corners of the earth.

Secondly, as all species rely upon the same base resources (land and NPP) for their sustenance, the marked expansion of some niches implies the contraction of others. Humans have usurped the evolutionary role as the allocation mechanism of base resources, and they have chosen to exercise this power on behalf of their 'chosen' species. The escalation of human-induced extinctions over the past ten thousand years is an indicator of this impact. This was the first phase of human extinctions, having little relation to the previous natural process. It made its first impact in the extinction of the large land mammals, whose requirements and ranges most closely matched those of humans.

To a large extent, the full impact of this new technology on biological diversity has been deferred by the dynamics of the process itself. The study of islands on earth indicates that species will react to human land alteration by concentrating in the remaining unconverted habitat. As the last refuges of unaltered habitat are converted, a geometrically increasing rate of extinctions is encountered.

This same process is now occurring at the global level, as it has previously occurred in cases of isolated habitats. Much of the productive parts of the northern hemisphere has been converted. At present the same process is working its way across the southern hemisphere, and into the vast tropical rain forest. Consequently, the next phase of the human-induced extinction process is commencing with the diffusion of human technology into these final refugia. The final spasm of extinctions at the completion of a ten-thousand-year-old process of technological diffusion is what we term the problem of 'biodiversity losses'.

Therefore, extinction is now a human choice process, deriving from this technological shift of ten thousand years ago. It bears little resemblance as a process to the constructive character of the natural force of extinction. In fact, it represents the human usurpation of the evolutionary role. Humans have exercised this power in most dramatic fashion by reshaping the earth in a very specific and specialised manner. This refashioning has afforded previously unavailable base resources to the human species (through its chosen surrogates), thus expanding its niche (and theirs); however, it has simultaneously deprived many other species of their own. The global problem of biological diversity losses has been generated by this new-found human power, but especially by the special way in which it has been exercised.

5 The Commons and the State: Regulating Overexploitation

5.1 INTRODUCTION

The overexploitation of renewable resources has been explained as the result of 'open access' conditions applied to a common resource. This analysis was initially developed by Gordon (1954), but then generally adopted by most other analysts of the overexploitation problem (see, e.g., Clark, 1976). Open access has been accepted as a fundamental explanation of the problem of overexploitation.

This overlooks the necessity of explaining why open-access conditions inhere in regard to certain resources and not in regard to others. For marine resources, the source of this distinction may be obvious, as some resources fall within domestic jurisdictions while others fall outside. The latter are available for all to access, while the former are not. However, in the case of terrestrial resources, this distinction does not apply, since all terrestrial resources fall within some state's jurisdiction. That is, for every terrestrial resource, there is an 'owner' – the state within whose borders the resource lies.

In addition, it is not apparent why open access is a useful concept when the same 'owner-state' capably regulates some resources (e.g. tin mines and tea plantations) while failing in regard to others (e.g. wildlife and forests). Under such circumstances it is probably better to view an open-access regime as the consequence of a societal choice, rather than the cause of collective failures. More fundamental forces are determining the owner-state's decision concerning which management regime to apply to a given resource or region.

For this reason, the open-access model of overexploitation cannot be a fundamental explanation of resource decline. Open-access regimes are as much an effect of the social processes causing resource decline as is the decline itself. They constitute one of several available routes for implementing a society's decision to disinvest in a particular resource.

116

Specifically, terrestrial open-access regimes are used to implement disinvestment decisions with regard to resources that are valuable enough to warrant harvesting, but not sufficiently worthy (as assets) to warrant the funds required for management of the harvest. In that case, the 'optimal' management for the particular resource might be practically none at all. Then the desired disinvestment in the resource occurs, without the necessity of incurring the costliness of managing the process.

The declines of many of the large land mammals in sub-Saharan Africa are illustrative of this manner of disinvestment. For example, four African range states (Sudan, Central African Republic, Tanzania and Zambia) accounted for the bulk of the decline in the populations of the African elephant that occurred over the past decade. These four states alone lost over half a million elephants (nearly half of the continental population) during the 1980s (ITRG, 1989). None of these elephant declines derived from management programmes, as it was illegal to take elephants in each of these states during this period (IUCN, 1985). Rather, it was a classic example of 'arm's-length' exploitation, as elephant populations were drawn down through largely unregulated access and subsequent government sales of confiscated tusks (Swanson, 1989a). These states could have better regulated the exploitation of this species, if it had been a high priority object to do so. Leaving the populations unmanaged indicated that other priorities were being ranked more highly by these overstretched governments, and equated with a tacit endorsement of the overexploitation that inevitably resulted.

The institution of the open-access regime is applied to a wide variety of resources in a large number of countries. The same sort of institutional arrangement that applied to the African elephant applies, in certain countries, to most wildlife (Swanson, 1991b), tropical forests (Repetto and Gillis, 1988) and various other diverse resources.

It is unsatisfactory to source this form of overexploitation in the existence of open access regimes. Under these circumstances open-access regimes are as much an effect of the base causes as is the overexploitation. It is singularly unhelpful to stop the analysis at this stage because, first, it fails to identify the true fundamental causes of overexploitation and, secondly, it fails to identify the general nature of the institutions required for regulating overexploitation (by focusing on the nature of the property regime rather than the general nature of the solution).

This chapter attempts to formulate the more fundamental nature of the open-access problem, and the general nature of the reforms required to resolve it. It demonstrates that a natural resource commons problem may be conceived of as the coincidence of three *joint asset investment problems*, involving the base resource, the biomass flowing from that base, and the requisite management for those resources. The most fundamental of the three is the last. Investment in management must occur before incentives will exist for investment in either of the other two natural assets.

However, management investments result only when there are incentives for them to occur, and the absence of such incentives is the essence of the open-access problem. This requires *institution-building*, i.e. investment in the creation of incentives for management. The irony is that, when these incentives are most needed, the incentives to develop them are least extant. Individuals in a common have no more incentive to invest in institution-building than they do in management. Then, this is a function that only government is able to fulfill.

One of the primary roles of government is to create such optimal incentive mechanisms. However, it is important to note that institutional reform of this nature will in general attract only economically 'optimal' levels of investment, i.e. investment funds in an amount equal to the returns that the reforms generate. This means that 'first best' institutions (from the perspective of natural resource management) may not exist, or even be targeted. Inferior natural resource regimes, such as open-access institutions, can result from the perceived inadequacy of the returns from investments in natural resource-related institutions.

Therefore, this chapter develops a model of the *bio-institutional framework* within which individuals operate in the context of natural resource exploitation. It is this framework which determines the incentives for individuals to invest in these natural resources, and thus determines their particular fates. It is the sovereign 'owner-state' that determines this framework for all of the terrestrial resources within its jurisdiction. Therefore, the overexploitation of natural resources must be the result of a choice (implicit or explicit) made by that owner-state. However, even nations operate under constraints, and it is this fact that ultimately generates the 'open-access' regime and the overexploitation that it implies; overexploitation is the result of the relative incentives for investment in various resources.

The chapter proceeds as follows. First, the differential impacts of different management regimes are illustrated in section 5.2, in the context of two very different regimes: open access and pure private property. Section 5.3 then demonstrates that the selection of the management regime is itself a construct of human choice. The state decides the nature of the management regime applicable to various resources, depending upon the ability of that resource to 'pay' for the costs of the institution-building. Section 5.4 generalises the points of the preceding two sections in the formulation of a general model of bio-institutional regulation.

The overarching objective for this chapter is to demonstrate that there is only one fundamental force for resource depletion: the incentives to invest (or not to invest) in that particular biological resource. In order to avoid depletion, biological resources require investment in their stocks (i.e. they must not be entirely disinvested). However, there are other, ancillary requirements equally necessary for their sustenance, e.g. base resources and management. When any of these required investments are withheld, depletion is the result.

Resource depletion occurs when there are no incentives to make these investments. One source of depletion is the lack of incentives to maintain stock of the resource; another source of depletion is the lack of incentives to supply the required base resources (land). A prominent source of depletion is the lack of incentives to supply the third necessary investment, in resource management. This results in the depletion of the resource by virtue of overexploitation.

However, the fundamental cause of this form of depletion is not the overexploitation but the lack of incentives to control it. These in turn derive from the general characteristics of the biological resource. When there are no incentives to invest in the biological resource, there are no incentives to invest in its management either. When this is the case, the result is an open-access regime and overexploitation. Even overexploitation is the consequence of the more fundamental problem: the absence of incentives for investment in the resource.

5.2 THE NATURE OF THE COMMONS PROBLEM: INSTITUTING INCENTIVES FOR INVESTMENT

In the natural resources literature, the *commons problem* has been conceptualised as a natural resource that is subject to mismanagement by reason of insufficient incentives to control individual exploitation. However, this conceptualisation confuses two distinct assets: the natural resource and its management. The solution to the commons problem requires incentive systems that generate optimal individual investments in both of these assets.

In fact, investment in management is a precondition to the creation of incentives for investment in the natural assets. It is the lack of incentives for investment in management which lie nearer the base of the commons problem. However, if incentives do not exist for investment in the natural resource, then it is unlikely that they will exist for investment in management, as the return to management flows through the same commons.

Therefore, the fundamental nature of the commons problem lies two stages removed, with the necessity of investments in the creation of incentive mechanisms to encourage investments in management. This is the need for *institution-building*, and it is at the core of the commons problem. It is the failure to invest in this underlying asset, institutions, that ultimately generates inadequate incentives for investment in the natural assets. It is the purpose of this section to demonstrate how and why this is the case.

The section first commences with an examination of the contrasting equilibria resulting under two very different management regimes: *open access* and *basic property rights*. It is demonstrated here that the fundamental differences between the two different regimes are the incentives for individual investments in management services, and the consequential impact which the resulting aggregate levels of management have on the incentives to invest in natural resources. This point is then generalised, when it is shown that the essence of an open-access regime is the societal determination not to invest in the management of a particular resource.

5.2.1   The Open-Access Regime: Institutionalised Incentives for Underinvestment

This section investigates the nature of the incentive scheme instituted by an open-access (OA) regime, in the context of an asset-based model of resource utilisation. In short, it demonstrates that the institution of an open-access regime implements a system of incentives that generates underinvestment by individuals in the natural resource.

These results may be derived in a simple model of renewable resource utilisation within the context of an OA regime. Consider a terrestrial species, with stock (x) and flow (y). This species resides within the jurisdiction of a given state (s), *on lands to which a large number of uncoordinated individuals* (i) *have access, or potential access*. This last provision constitutes the core criterion of an OA regime. 'Open access' means precisely what it says: access is open to any individual with the incentives to use the resource.

What is the effect of this regime on the particular resource (e.g. a species) to which it is applied? This regime alone will determine the incentives for investment in the species and its habitat. In short, the application of an OA regime will create disincentives for investment in the resource, which are equivalent to forces for the depletion of the resource.

This is so because human investment is now the crucial factor determining the viability of a species. That is, the human species now holds the 'evolutionary prerogative': the power to determine the allocation of life-sustaining resources between the various life forms on earth. If society chooses not to invest in a particular species, this is now equivalent to a decision to disinvest in the species, although the particular route to disinvestment may be one of several. It all depends on which of the required resources (that are withheld) is the first 'to bite'.

The essence of the decision model is that human societies will make decisions concerning allocations of required resources to particular species, in the course of choosing a portfolio of productive assets. If the resource is to be included within this portfolio, then this requires a decision not to disinvest in the resource stock, and also a decision to invest in the ancillary resources required for its sustenance. That is, in order to avoid depletion, a given stock of the species must be left at the end of each period (x_t) in order to allow the species to regenerate itself, and in addition sufficient lands (R_t) must be left in a form which the species finds conducive for its survival (e.g. the trees must remain intact if it is a forest species). If either of these conditions does not inhere, then the species will experience disinvestment and ultimately extinction (through either stock disinvestment or land use conversion).

In equation (5.1), the incentives for a human society to provide these resources to a given species are examined within the context of an open-access regime. The essence of an OA regime is that the individuals concerned consider only the benefits to be realised from the utilisation of only two *joint assets* – the species stock (x) and the land (R). The other investment decision is made for them; there is no management in this commons. Access to the resource is available to all (n) individuals with the capacity to harvest there.

These two jointly-held assets then yield individual benefits through their impact on both resource growth $(H(x,R))$ and the cost of harvesting $(c(R))$. The decision problem concerns the determination of the individual's optimal rate of harvesting (y_i) and the individual's optimal rate of land allocation to this species (R_i) (by, for example, not harvesting the timber within the forest). These are, in effect, the equilibrium levels of individual investments in the natural assets resulting under the prevailing management regime.

Incentives for Individual Investments Under an Open Access Regime
(Individual Choice Problem Regarding Investment in Joint Assets x *&* R)

$$\text{Max}_{y_i, R_i} \prod_i^{x,R} (y_i, R_i; x, R) \equiv \int_0^\infty e^{-rt}\left[p(y)y_i - c(R)y_i - r\rho_R R_i \right] dt \quad (5.1)$$

subject to $\dot{x} = H(x,R) - y$

where:

$\Pi_i^{x,R}$ ≡ individual benefit from the use of joint assets x, R

$p(y)$ ≡ demand for goods of species y

$c(R)$ ≡ unit cost of harvesting y

y_i ≡ harvest of species y by individual i

R_i ≡ lands unmodified by individual i suitable for use by species

ρ_R ≡ price (opportunity cost) of unit of unconverted land

$H(x,R)$ ≡ growth function of species

r ≡ cost of capital applicable to individuals in community

n ≡ the number of individuals with access to the common

In order to solve this decision problem, it is necessary to specify the nature of the interaction between the various agents with access to the assets. The sort of conjectures which best describe interaction in the context of an open-access regime are known as Open Loop Nash Equilibrium (OLNE) assumptions. Open-loop conjectures imply that the only information available or relevant is that arising in the current period. This applies well to an open-access situation because there is little opportunity for a feedback-based equilibrium when there is free entry. Enhanced cooperation within the environment will generate additional profits and so more entry, thus returning the group to its original, non-cooperative state. In addition, Nash assumptions (that individuals do not respond to other individuals' strategies) apply well in any situation where there are large numbers of players, because it is difficult to develop more sophisticated strategies in this context.

Given this manner of interaction, it is possible to derive the following equilibrium 'investment conditions' for the individuals accessing the common. These investment levels that arise under an open-access regime are termed (x^{OA}, R^{OA}).

Individual Investment in Diverse Resources under Open Access (OA) Regime (OLNE)

$$y_i^{OA} : \frac{\lambda}{n} = p(y)\left[1 + \frac{1}{\varepsilon_D n}\right] - c \tag{5.2}$$

where: λ ≡ the value of marginal relaxation of the stock constraint (the rental value of the resource stock)

Stock Investments

$$x^{OA} : \frac{\lambda}{n} \cdot H'_x + \frac{\dot{\lambda}}{n} = r\lambda \tag{5.3}$$

Base Resource Investments

$$R_i^{OA} : \frac{\lambda}{n} \cdot H'_{R_i} - \frac{C'_{R_i} y}{n} = r\rho R \qquad (5.4)$$

Consider the conditions determining resulting individual investments in stocks (5.3) and in base resources (5.4). In each, the equilibrium condition may be interpreted as balancing the returns from investments in this diverse resource against the returns available on the marginal investments within the economy. In these two equations, the left-hand side of each may be interpreted as representing the anticipated return from investing in this diverse resource; the right-hand side of each equation represents the anticipated return from investing in other assets within the economy. In each case, there are various discount factors applied when deciding the level of investment in a natural resourse, on account of the nature of the management regime to which it is subject (i.e. open access).

Equation (5.2) establishes the definition of resource 'rent' under an open-access regime. If there are large numbers of suppliers in the product market (i.e. $n \to \infty$), this implies that $\lambda \to (p - c)$. Resource rent is then equal to the residual from price over input costs.

The first disincentive to investment under open access derives from the 'averaged return' received from such investments. This is reflected in the term on the left-hand side of equation (5.4) which contains the expression (y/n). Here, the 'individual return' is equal to the joint yield divided by the number of harvesters. Individual investments flow into the natural environment, but returns flow uniformly back out of the same. Open-access regimes provide the ability, and hence the incentive, to free-ride.

The second disincentive to investment inhering under an open-access regime derives from the failure to manage entry into the commons. For this reason, the value of λ is divided by the number of harvesters, on the left-hand side of these equations. This represents the expected value of an investment in the diverse resource, given that the probability of future appropriation of the return from that investment is the same for all harvesters (the one investing and all others).

Of course, it is also possible to predict the number of entrants and the expected average returns from investments. The expected returns from investments will be eroded via entry to that point where the average return is equal to the cost of access. Given low costs of access (ready transport between habitat and population centre), the expected returns on investments in the common must be approximately zero. In that case, the incentives to invest in these joint assets will also be non-existent.

Therefore, the institution of an open-access regime clearly creates a system of incentives for disinvestment in the assets subject to its management. All biological resources, irrespective of their relative capacities for growth (H'_x and H'_R), are rendered non-viable investments by the institution of an open-access regime.

5.2.2   The Introduction of Property Rights Institutions – An Example

This section is intended to demonstrate the nature of the incentives inhering under a very different form of management regime: individualised property rights. In this system there are no joint assets whatsoever; each individual deals only with assets that are wholly 'individualised'. This of course creates a very different system of incentives for investment; the asset managed under such a regime receives investments that are determined much more by its own characteristics rather than the regime's. This is the rationale underlying the familiar prescription of 'property rights regimes' for the resolution of commons problems (Demsetz, 1967).

However, this section is also intended to demonstrate that the answer to commons problems is more complicated than the simple prescription of 'well-defined property rights', because the institution itself is not a unitary concept. Moving away from the regime of open access implies the movement across a spectrum of possibilities (not a movement to a single point). These possibilities are defined by the amounts of funding that the state is willing to invest in the construction (and maintenance) of an institution other than open access for the resource, and the resulting amounts that individuals are willing to invest (under this institution and investment) in the management of the resource. It is the sum of all of these investments (state and individual) that determines the nature of the management regime governing a resource, not its name.

For example, a 'parks and protected areas' movement initiated by the International Union for the Conservation of Nature (IUCN) about thirty years ago has resulted in the designation of vast tracts of land as 'national parks' throughout the developing world. These were the results of a previous international attempt at institution-building for natural resource conservation. However, these designations have had little or no impact on natural resource exploitation in the areas they concern; they are little more than 'paper parks'. This is because the adoption of legislation denominating land a national park, and the creation of maps listing these newly-designated parks, without more, has no impact on state and individual incentives to invest in the management of these lands. For the most part, the new designation did little or nothing to alter the exploitation situation away from the previously prevailing conditions, often open access.

True institution-building acts to modify the framework within which individuals operate, and hence alters the existing incentive structure regarding individual investments in management and, consequently, natural resources. The chain of causation in the process of institution-building is as follows. First, the state acts to alter the incentive framework within which individuals make decisions regarding investments in management. Then, individual levels of investments in natural resources *react* to their own investments in management. This section demonstrates the process by which the system of incentives under OA might be revised with an initial act of institution-building.

Assume that the institution-building commences when a state decides, for some reason, that it would be useful to regulate more efficiently some natural resources currently under an open-access regime. At this juncture, however, the state is assumed to be unwilling to make a commitment to substantial and ongoing investments in management. The baseline approach to the commons problem might be the institution of a very basic property rights regime. In this situation the state invests only to the extent necessary to undertake an 'authoritative' allocation of the base resource to individuals (with an assurance of sanctions for apprehended interlopers), while leaving the ongoing implementation of the system to those same individuals.

In this instance the sole investments in the day-to-day monitoring of the property allocation (i.e. its 'management') will be those made by the individuals in the course of defending the rights they have received. It is very unlikely that this will always be the most efficient form of institution-building, but it is used here in order to distinguish between state and individual involvement in monitoring in order to present the simplest possible case of institution-building. Here, the cost-effectiveness of the state's intervention will not be considered (as this is the subject of section 5.3); the relevant issue here is how this smallest-possible state intervention impacts upon individual incentives for investment in management.

The introduction of such a basic property-rights institution may be conceptualised as the creation of a new strategy for value appropriation. It has the effect of giving some individuals the idea of investing in the exclusion of others from some individual segment of the base resource (this will now be termed R_i). The biomass associated with that base resource will be similarly individualised (and thus termed x_i) to the extent that others are excluded by individual investments.

Investments in management will be assumed to be of the nature of fencebuilding. The more that is invested in exclusion, the higher is the fence, and the more potential entrants will be disqualified. However, such investments can never be effective with certainty. The distribution of potential entrants will be assumed to cover a wide range of fence-climbing abilities and motivations, and thus exclusionary investment would need to go to infinity for all potential entrants to be excluded. Increasing investment will keep an increasing proportion of the entrants out of the area, but at a decreasing rate.

Under these assumptions, the institution of property rights has a very specific sort of impact, by creating incentives for individual investments in management (M_i) which thereby create incentives for individual investments in the other joint assets: the base resource and the biomass. The individual resource management problem, (5.1) above, can now be reformulated as follows, under the assumptions of this very basic property rights regime. The important point to recall is that under this regime there are no longer any joint assets.

Incentives for Individual Investments under a Property-Rights Regime (PR) – Individual Optimisation Problem regarding investment with no joint assets

$$\underset{y_i, R_i, M_i}{\text{Max}} \prod_i^{PR} (y_i, R_i, M_i) \equiv$$

$$\int_0^\infty e^{-rt} \left[p(y)y_i m_i(M_i) - c(R_i)y_i m_i(M_i) - r\rho_R R_i - r\rho_M M_i \right] dt \quad (5.5)$$

$$subject \ to : \dot{x}_i = H(x_i, R_i) - y_i$$

where:

$\prod_i^{PR}$ ≡ the benefits to individual i from investment in allocated assets

m_i ≡ the proportion of the flow appropriated by the investor i

y_i ≡ the proportion flow of species y from base resource unit i

R_i ≡ the amount of base resource unit i devoted to production of species y

x_i ≡ the stock of species y associated with base resource unit i

M_i ≡ the amount of investment in the management of base resource unit i

This formulation of the problem states that the construction of such a basic property-rights regime involves the allocation of 'islands' of the base resource to individuals, on which the stock and flow of biomass is then perceived to be segregable from surrounding lands. The impact of this allocation is to create the impression that there are no longer any joint assets. That is, other individuals' investments no longer have any perceived effect upon the individual's return. This belief creates incentives for individual investments in the management of these individualised natural resources.

Equation (5.6) states that individuals within this very basic property rights regime will now invest in management services corresponding to the allocated base resource unit (i). That is, individuals will invest in 'building fences' around their allocated assets to the extent that these investments generate competitive returns. The state's limited effort at institution-building has thus introduced management into the habitat for the first time.

Management Services Investments under Basic Property Rights Regime (PR)

$$M_i^{PR} : \frac{m_i' \lambda y_i}{m_i} = r\rho M \tag{5.6}$$

where $\dfrac{\lambda}{m_i} \to p - c,$ if the market in which y is sold is competitive

Management is the key to providing the possibility for natural resource investments. The individual investments in management that result under this basic property rights regime provide the framework within which individual investments will then be made in the natural resources. There is now in effect an incentive system which gears individual returns to individual investments in the natural resources. Therefore, there is now a great higher optimal level of individual investments in the natural resources *given the higher level of investment in management.*

Natural Resource Investments – Base Resource and Stock (PR)

$$R_i^{PR} : \lambda H'_{R_i} - c'_{R_i} y_i m_i = r\rho R$$

$$x_i^{PR} : \lambda H'_{x_i} + \dot{\lambda} = r\lambda \tag{5.7}$$

where $\lambda = m_i(p - c)$ in a competitive industry

The individual investments in management have the effect of overhauling the investment incentives regarding the natural resources. Comparing equations (5.3) and (5.4) with the equations in (5.7) indicates that the 'free-rider' effect is now reduced by reason of investments in determining the level of flow appropriation (m). That is, investments in the base resource no longer yield an average return (y/n) but an *individualised* return (y_i m), dependent upon the individual's investment in management. The same is true with regard to investments in the stock of the species; forebearance in harvesting results in an impact on base resource unit i alone, channelling the returns from investment to the investor.

The 'free-access' effect under an open-access regime is also removed through management. The expected rents from the resource (λ) under open-access had been minimal; there is pressure on rents to go to zero under open-access if access is readily available. However, under a basic property rights regime, λ is now determined by the individual's management spending. That is, the unit rental value will be in equilibrium between the force of potential entrants and the force of the investor's exclusionary investments.

Therefore, under the property-rights regime, there is the prospect for investments in the biological resources. The level of these investments will now be determined, at least in part, by the individual characteristics of those resources (H'_x and H'_R).

Underlying all of the differences in these two equilibria (OA and PR) is the government's choice of institutional framework. That is, all of these investments, in management and consequently in the subject natural assets, are the consequence of the state's initial investment in altering the bio-institutional framework, i.e. in altering the individual's perception of the incentive system applying to management. Prior to an analysis of the state's choice of the institution to apply to different resources, it is important to generalise the analysis, so that the entire range of institutions can be considered.

5.2.3   The Fundamental Nature of the Commons Problem: Under-investment in Management

The commons problem falls generally within that group of problems which concern the optimal management of 'joint assets'. This is the general nature of many different types of relationships: firms, contracts, distribution agreements (Grossman and Hart, 1987; Hart and Moore, 1990; Williamson, 1986). The classic commons problem is simply the instance where one of the joint assets concerned is a natural resource.

It is fundamentally important to distinguish between the tangible assets that are the context of the relationship and their management, which is another dimension. That is, in regard to the commons problem, it is important to distinguish between natural resource utilisation and natural resource management, as they are distinct 'joint assets' in the context.

In many relations, it is easy to distinguish between the context and the incentive system. In particular, where the relationships are purely 'vertical' (such as the passage of goods through a chain of distribution), there is no point in time when the tangible asset is actually held in common. Then, the only joint asset is the incentive system applying to the various parties in the relationship.

For example, consider an insurance company that develops an insurance contract and then distributes it through a network of independent sales and service agents. The joint asset between company and distributor is not the insurance policy (which is the sole property of one party in the chain of distribution at each point in time), but the system of incentives to generate optimal investments by both. An institution must be developed within this purely vertical system that will create incentives for agents at both levels to invest optimally in the maximisation of the value of the policy. This system must generate incentives for the insuror to invest in the creation of the value-maximising terms in the policy and must also generate incentives for the retailer to invest in the value-maximising sales and service scheme. It is this *system of incentives to individual investments* ('management') that the two parties hold jointly, not the contract itself.

This indicates why the natural-resource commons problem involves more joint assets than have usually been considered. In this case there is both a *horizontal* dimension (the natural resources) and a *vertical* one (their management). That is, unlike the case of the purely vertical relationship, the commons involves joint production in the context of common tangible assets. However, just as in the case of the purely vertical relationship, there remains another joint asset that is distinct from the tangible assets, i.e. the incentive system (or 'management').

Management is an entirely distinct asset which itself requires investment: it is not another name for investment in natural assets. This means that the commons problem may be conceptualised as production in the context of three distinct joint assets: *the base resource, its flow of biomass, and their 'management'*.

The examples of the previous sections demonstrated how the most fundamental joint asset in the context of the commons is the provision of the optimal incentive structure for investments in the natural assets ('management'). Investments in management were a precondition, in sections 5.1 and 5.2, to the development of incentives for investments in the natural resources.

This is depicted in equation (5.8). In the generalised commons problem there are now three joint assets (x, R, M), and it is the third (M) that drives the resolution of the commons problem. Aggregate investment (M) determines all individual incentives for investment in management (M_i), in base resources (R_i) and in the resource stock level $(x(y_i))$.

Generalised Commons Problem – Individual Optimisation Problem
(Regarding Investment in Joint Assets x, R and M)

$$\underset{y_i, R_i, M_i}{\text{Max}} \; \prod_i^{x,R,M} (R_i, M_i; x, R, M) \tag{5.8}$$

$$\begin{aligned}
\textit{subject to} \quad \dot{x} &= H(x,R) - y \\
y_i &= y_i(M) \\
R_i &= R_i(M) \\
M_i &= M_i(M)
\end{aligned}$$

The generalised commons problem derives from the existence of these three joint assets; however, the fundamental nature of the commons problem is indicated by the underlying importance of aggregate investment in management institutions for the creation of individual incentives for investment in the other assets. Investment in management, M, underpins individual decision-making with regard to investment in the base resource, $R_i(M)$, and with regard to investment in the stock of the individual species, $y_i(M)$.

At the core, then, the commons problem concerns the incentives which exist for investing in developing a management regime. This is a tertiary layer of incentives: the incentives to develop an incentive system for the natural assets. The two management regimes considered at the outset of this chapter (OA and PR) are discrete examples of the forms of incentive systems which may exist; however, there is a continuum of possible systems. The range of possibilities concerning this most fundamental layer of incentives are termed *institutions*, and this core problem of the commons will be termed *institution-building*.

The fundamental problem of the commons is that, in the absence of already-existing investments in institution-building, there are insufficient individual incentives for the commencement of such investments. This is because the same incentive system applies to investment in the underlying joint assets (of management services and institution-building) as applies to the other joint (natural) assets. In order to demonstrate this, consider these underlying incentive systems as a single combined asset, termed management. From the perspective of the interacting individual harvesters, this is in fact the case; management investments will either be *direct* (expenditures on monitoring and appropriation) or *indirect* (expenditures on the creation of an incentive system for such investments). The allocation of management spending will depend on the relative return to direct versus indirect spending.

In either case, management is an intangible asset that yields its return when combined with other, tangible assets. In the case of the commons problem, management is combined with natural assets, and its return is channelled back through them. Therefore, starting from the initial position of little or no aggregate investment in management, there is little individual incentive to commence investing optimally.

This may be demonstrated by the introduction of the fundamental joint asset of management into the individual's decision problem set out in (5.1) above; this is done below in equation (5.9). Now the individual must decide how much to invest in initial 'fencebuilding' (management), under conditions of open access. Under these initial conditions of little or no aggregate investment in management, individuals will only invest in this core asset in order to equate marginal private benefits with marginal private costs; that is, the societal benefits from individual investments in these base incentive mechanisms will be ignored in making the investment decision.

Management Services Investments by Individual under Commons Regime (CP) (OLNE)

$$Since \ \prod_i^{x,R,M} \equiv \prod_i^{x,R}(M) - r\rho_M M_i$$

$$\longrightarrow \ M_i^{CP} : \frac{\partial \prod_i^{x,R}}{\partial M_i} = r\rho_M \tag{5.9}$$

where: $\rho_M \equiv$ is the price of a unit of management services
$\quad \ M_i \equiv$ is the individual investment in management services

Equation (5.9) implies underinvestment in management services, precisely because management is a joint asset. Ironically, when the underlying incentive systems are weakest (and institution-building would potentially yield its greatest aggregate return), the individual incentives to invest in institution-building are also at their weakest.

This discussion may be used to illuminate the special nature of the case of OA regime. In particular, it will be used to demonstrate the *polar* nature of an open-access regime along the general range of possible institutions. OA is the system of incentives that prevails with regard to natural resources when there is little or no aggregate investment in the management of these assets. That is, the OA regime is a special case of the three-asset model discussed here, but with aggregate investment in the base assets (management and institution-building) very near zero. Then, the observed result will be aggregate investments in natural resources going to zero, and consequent overexploitation.

The Nature of an Open-Access Regime – Non-investment in Management

If $M = 0$,

$$\longrightarrow \prod_i^{x,R,M} = \prod_i^{x,R}$$
$$\longrightarrow x\,(M=0) = x^{OA}$$
$$R\,(M=0) = R^{OA} \tag{5.10}$$

This discussion also indicates why open access regimes are very stable institutions – in the absence of governmental intervention. If the initial position is one of low aggregate investment in management institutions, there is virtually no individual incentive to commence such investments. This is because the returns from investments in management services will then be channelled through a joint asset regulated under an open-access regime, which means that the returns will be very poorly directed. Under these conditions, the incentives for *individual investments* in management will be virtually non-existent. This means that an open-access regime, once prevailing (and without intervention), will remain a stable equilibrium.

This section has depicted the nature of the commons problem as one of underinvestment in joint assets. The joint assets implied by the commons problem are three: the base resource (land), the resident biomass (species) and the management of these natural assets. The most fundamental of the three is the last. Investments in management determine the incentives for investment in the other joint resources. However, as the returns from management are channelled through the commons, the same undeveloped individual incentives inhere for management as for the other (natural) assets. Furthermore, since the reform of this incentive system is also a management function, it also requires resource commitments for its resolution. The core problem of the commons is that, once institutions are in existence which generate low levels of individual investment in management, it is improbable that new institutions will be forthcoming from those investments alone.

5.2.4 The General Nature of the Solution to the Commons Problem: Investments in Institution-Building

To this point this section has demonstrated the fundamental nature of the commons problem, which is the inadequacy of investment in the base joint asset: management. When some amount of investment in *management* occurs, there are incentives for investment in the other joint assets of the commons (i.e. the natural resources). With inadequate investments in management, regimes akin to open access are instituted, and there are few incentives for investment in natural resources. This is how overexploitation occurs in the context of the commons problem.

This basic example has also indicated the general nature of the solution to the commons problem. The general solution is the modification of the incentive structure so that investments in management will occur, a type of management spending termed *institution-building* in this section.

An example of institution-building was provided in section 5.2.2 in the context of a basic property-rights institution. It is important to note that, although the property-rights form of institution is one possible means for addressing the core problem, it is not the sole means of doing so. That is, the movement along the spectrum of possible institutions away from OA regimes implies movements toward generally increased aggregate levels of investment in management, and not necessarily movements toward property right institutions. Therefore, the simplest statement of the general nature of the solution to the commons problem is: *increased aggregate levels of investment in management.*

The importance of the distinction lies in the fact that there is no *a priori* basis for believing that property-right institutions are the first-best approach to the solution of the commons problem. The determinative characteristic possessed by property-rights institutions is that they minimise state investment (M_s) in management relative to institution-building. That is, property-right institutions involve a strict division of labour between private and public sectors with regard to direct and indirect investments in management. Again, there is no *a priori* basis upon which this division of responsibility might be deemed optimal.

Property-rights institutions are an attempt to institute a 'vertical' system of incentives through operation solely on a 'horizontal' plane. That is, they are providing management through the simple expedient of the subdivision of the resource requiring management. As mentioned above, this unnecessarily confuses the two distinct assets, management and natural resources. It is very possible that optimal joint management will not require interference with the manner in which the other joint asset is held.

It is also based on the assumption that these joint assets may be beneficially segregated. That is, property-rights systems create the artifice of individualised 'islands' in an actual continuum of base resource and biomass. However, externalities will necessarily persist on account of the existence of joint assets, resulting in inefficiencies in investment across the system (as there is leakage of the impact of investments across boundaries).

Also, this manner of institution-building will not be compatible with all forms of biomass. Some forms will not persist under conditions of individualised exclusionary investments, e.g. migratory species. Some forms will not easily be appropriated by individualised exclusionary investments, e.g. the avian species. There will be a necessary loss of a significant portion of life-form diversity, and some particularly valuable species, by the exclusive reliance upon property rights institutions.

These two broad reasons, *the persistence of externalities and the exclusion of options*, indicate why property-rights institutions would not always and everywhere be the optimal direction for institutional development in the commons. The fundamental nature of the commons problem is the need for the development of incentives for investments in management. The general nature of the solution is investment in institution-building and management in order to move upwards along the spectrum of available incentive systems.

5.3   OPTIMAL INVESTMENTS IN INSTITUTION-BUILDING IN THE COMMONS

The models of the previous section suggest the nature of the solution to the commons problem: this is the movement toward an institution that generates optimal investment in the management of joint (natural) assets. This section concerns the nature of this institution-building process, i.e. the incentives for investment in institution-building. In particular, it addresses the question concerning why one form of institution is applied to some natural resources, and very different forms of institutions applied to others.

This enquiry requires the identification of the agency making decisions regarding institution-building with regard to any particular commons. This is not difficult to do. Since there exists an 'owner-state' that claims sovereignty for every parcel of land, there then exists a regulator *de jure* of the incentive systems extant in every commons. Yet, in practice, it appears that in many cases the regulator abdicates this authority, leaving individuals and resources to reach ostensibly unregulated equilibria. This result is then denoted the inefficient overexploitation of natural resources.

However, this assumes that institution-building is a costless activity. In fact, the development of institutions requires the investment of resources which are thereby disallowed to other sectors of the economy. Institutional development regarding natural resources must compete for capital, just as does any other form of development (health, education, housing, etc.) This constraint on investment capital is the fundamental reason for the existence of inadequate management institutions, and hence overexploitation. That is, the absence of investments regarding certain natural resources usually corresponds to an official decision that these resources are incapable of generating a return warranting that investment. This section investigates this hypothesis, and demonstrates how states make decisions regarding *optimal* investments in natural-resource institution-building.

5.3.1 The Role of the State in Institution-Building

Institutions may be developed through either *horizontal* or *vertical* methods. Horizontal methods constitute attempts to recognise and internalise interdependence among those individuals facing the same bio-institutional framework. In this situation, institutions may evolve because individuals include within their choice problems not only environmental variables but also the choices of other individuals. Choices are then made, in part, in order to influence the choices of others, through various reward-and-punishment strategies. Vertical methods for institutional creation instead involve attempts by a third party (i.e. some individual or organisation facing a distinct optimisation problem) to reshape the bio-institutional framework (and hence the decision-making process) of the interdependent group. In this situation, the third party includes within its decision problem the choices made by the group in reaction to its choices. Vertical institution-building is similar to a Stackelberg-type of decision problem.

This section focuses upon the latter, vertical form of institution-building. There is a substantial literature developing around the idea of horizontal institution-building (see e.g. Olstrom, 1990). However, it has little practical importance for most terrestrial habitats, for two reasons. First, there is no remaining terrestrial habitat (with the possible exception of some segments of Antarctica) which remains unclaimed territory. This means that a specific state has exclusive authority for developing institutions with regard to every parcel of land on earth. Therefore, the vertical form of the institution already exists *de jure* across the face of the globe; the important question is why there is so little investment in this institution in so many places in the world, resulting in its virtual non-implementation *de facto*.

The other reason why horizontal institution-building is of little practical importance for terrestrial resources is technological. A technological shift in transport and communications has occurred across the globe over the past few hundred years. This has rendered every piece of terrestrial habitat vulnerable to the forces of open access. There is no place on earth now that is unapproachable over land. Vast urban populations may reach the frontiers in a matter of hours. With this omnipresent pool of potential entrants, there is no capacity for interdependence and recognition to evolve institutions over time. If the number of interacting, and potentially interacting, individuals is very large, then the transactions costs of such collective management is usually too great for such 'horizontal' contacts to evolve. Then it is necessary to rely more on 'hierarchy' for the provision of management solutions.

This scenario fits most natural resource exploitation situations. The key elements of the problem are the free and easy access of most of the nation's population. As transportation to the frontier improves with the development of roads and diffusion of motor transport, the pool of potential entrants into any natural habitat becomes as large as the population of the nation. Most natural-habitat exploitation problems occur on frontiers by reason of new colonists following new roads, not by reason of existing populations overexploiting their own lands (Southgate and Pearce, 1988). With such large numbers of foreign entrants, there is little prospect for the evolution of common resource management on a horizontal basis.

Therefore, it is the role of the state to provide for the optimal solution of these multiple joint-asset problems. The fact that the state claims juris-diction to the commons indicates that the vertical form of regulation exists *de jure*, if not *de facto*. In the large numbers context, there is little choice but to look to vertically-instituted incentive mechanisms. The core issue is: Why don't societies invest in the development of these institutions?

5.3.2 The Nature of State Institution-Building in the Commons: Working Through The Bio-Institutional Framework

The purpose of state institution-building in the commons is the generation of investments in the management of natural resources. It may invest resources in management itself, or invest in the creation of incentive mechanisms which generate investments in these assets. In either case, the state intervenes by investing in the management of joint assets, M_s. This investment operates both *directly*, through the public provision of management and monitoring, and *indirectly*, through the public creation of private incentives for individual investments in the same $M_i(M_s)$. It is the sum result of these two interventions which determines the aggregate societal level of investment in management, M. These aggregate levels of investment in management then determine the incentives for individual investments in the base natural resources (x, R).

State Investment in Management of the Commons

$$M^*(M_S^*): \underset{M_S}{\text{Max}} \int_0^\infty e^{-rt} \left(\left[\sum_1^n \prod_i^{x,R,M} (y_i, R_i, M_i; x, R, M) \right] - r\rho_M M_S \right) dt \qquad (5.11)$$

subject to :

$$x = x(M) \equiv \sum_1^n x_i(M)$$

$$R = R(M) \equiv \sum_1^n R_i(M)$$

$$M = M(M_S) \equiv \sum_1^n M_i(M_S^I) + M_S^D$$

$$\dot{x} = H(x, R) - y$$

The state's decision problem in (5.11) is of the Stackelberg type. The state is able to choose its social optimum regarding natural resource investment only indirectly, by operating through the reaction functions of the individuals in the society (the first two constraints). It does this by investments which alter the bio-institutional framework (the latter two constraints) within which individuals make their choices. The state invests in management both indirectly (M_S^I) and directly (M_S^D) in order to achieve the beneficial movements in the social–biological equilibrium that result from these investments.

Therefore, the state takes action with regard to the management of natural resources only through the alteration of the framework within which individual decisions are made with regard to these resources. Equation (5.11) demonstrates the interaction of the government sector, the private sector and the bio-institutional sector in the determination of the resulting social –biological equilibrium regarding these natural resources. This model is further developed in section 5.4. At this juncture, the focus is on the nature of the state's objective regarding commons management.

5.3.3   The State's Objective in Commons Management

The depiction of the state's decision problem in equation (5.11) implies that there exists a trade-off between natural resource management and alternative investments. The resources allocated to institution-building (for this natural resource) might be allocated instead to other investments in the economy. Thus, a state with sole jurisdiction over a common resource has two distinct problems to solve: the problem of first-best natural resource investments and the problem of optimal social investments. To some extent, the two objectives do not coincide on account of the opportunity costs implied in the former.

Consider initially the problem of first-best natural resource investments in isolation. First-best natural resource investment concerns only the nature of the 'first-best' solution to the natural resource portion of the commons problem. This is the problem of determining the first-best aggregate levels of investment in the natural resources (x^{fb}, R^{fb}), *assuming that the costliness of inducing these investments is zero*. Equation (5.12) defines the level of these first-best investments for the three joint assets, given that the cost of management investments (ρ_M) is zero.

A Subsidiary Problem of the Commons: First-Best Natural Resource Management (FB)

$$x^{fb}, R^{fb}, M^{fb} : \underset{x(y), R, M}{\text{Max}} \int_0^\infty e^{-rt} \left[\sum_{i=1}^n \prod_i^{x, R, M | \rho_M = 0} (y_i, R_i, M_i; x, R, M) \right] dt$$

$$\text{subject to} : \dot{x} = H(x; R) - y \tag{5.12}$$

However, the overarching problem of the commons is to determine and to implement the optimal aggregate level of investment in all three of these joint assets: *base resource, biomass and management*. The optimal social investment problem subsumes the natural resource management problem. There is no reason to assume that a state would choose to invest in institution-building to an extent that deprives more worthwhile investments of funding.

The State's objective is to invest *optimally* in the pursuit of first-best institutions. That is, the state acts by moving the system closer to the 'first-best targets' provided in equation (5.12), and thereby removing the costliness of less than first-best institutions. However, investments in institution-building have intrinsic costliness, as they allocate scarce societal investment funds to the management of these particular natural resources.

The state faces a trade-off between investing in achieving first-best institutions regarding these natural resources, and investing in other assets.

Therefore, an alternative way of depicting the *state institution-building problem* is to focus instead on the minimisation of the sum of these two types of costliness: *natural resource institutional inefficiencies and natural resource institution-building costliness.* This is depicted in (5.13) in two terms: the first represents the costliness arising from the 'distance' between first-best and existing institutions, the second represents the costliness of investments to reduce this distance.

Optimal State Investment in Management of the Commons (Cost Minimisation)

$M^*(M_s^*)$:

$$\underset{M(M_S)}{\text{Min}} \int_0^\infty e^{-rt}\left[\sum_1^n\left[\prod_i^{|\rho_M=0}(y_i^{fb},R_i^{fb},M_i^{fb};x^{fb},R^{fb},M^{fb})\right.\right.$$

$$\left.\left. -\prod_i^{|\rho_M\neq0}(y_i,R_i,M_i;x,R,M)\right]-r\rho_M M_S\right]dt$$

subject to: $\dot{x}=H(x,R)-y$ \hfill (5.13)

First-best incentive structures are a target, but not usually a feasible one. The internalisation of externalities requires continuing and ongoing investment in management. The price of that management is unlikely to be zero in any situation. The society faces a trade-off between investments in providing this management, and investments in other assets in that society.

Therefore, the objective of the state in commons management is to move the bio-institutional framework along the institution spectrum (via investments) toward a first-best incentive system. However, given the intrinsic costliness of institutional investments, it is unlikely that first-best will ever be the ultimate target. This investment is constrained, as are all investments, by the opportunity cost of foregoing investments in other assets in the society.

5.3.4 Optimal Institutions Generally: Providing Optimal Incentive Systems

One role of the state is to provide *optimal institutions*, i.e. optimal incentive mechanisms for organising society, because these are public goods that will not otherwise be provided. However, this does not imply that it will necessarily be optimal to provide first-best incentive mechanisms with respect to each and every resource. This is because the provision of institutions itself entails costliness. The 'optimal' institution is the one that balances the benefits from enhanced incentives with the costliness of institutional modifications.

This constitutes a revision to the assumptions incorporated within the analyses of vertical relationships (e.g. the 'principal–agent' (P–A) sort of incentive problem), in order to take into account the costliness of institutional modifications. Under the P–A problem, the principal's object has been assumed to be the installation of first-best incentive mechanisms at minimum costliness, subject to the constraint of maintaining all agents within the principal's jurisdiction (Grossman and Hart, 1987). In the context of the commons, the analogous objective would imply that it is the state's role to achieve first-best investment incentives at the minimum possible cost.

Such a rendition of the model eliminates the possibility that the societal optimum might lie with investing society's resources in some endeavour other than the refinement of these incentives. The 'optimal institution' model provides for this possibility. There is an express trade-off between the costs and benefits of commons management implicit in this model.

The *optimal institution* is the one which balances the gains from governmental expenditures on incentive modifications against the costliness of those expenditures. Because the institutional change is a collective good, the correct assessment of optimality considers the sum of the individual benefits. (Samuelson, 1954). This condition can be derived by differentiation of either equation (5.11) or (5.13).

Optimal Social Investment in Institution Building

$$M^* : \int_0^\infty e^{-rt} \left(\sum_{i=1}^n \frac{\partial \prod_i (M)}{\partial M} \right) dt = r\rho_M \qquad (5.14)$$

Therefore, the role of the state in commons management is to invest optimally in the creation of the correct individual incentives to investment, and this requires the consideration of the aggregate societal impacts of any investment in management. However, this does not imply that first-best incentive systems will necessarily result, or even be targeted. It all depends on the capacity of the institution-building process to generate a competitive return.

5.4 REGULATING OVEREXPLOITATION: THE GENERAL MODEL OF BIO-INSTITUTIONAL REGULATION

This section builds upon the arguments of the previous sections in order to develop a generalised model of regulation in the commons. The *general model of bio-institutional regulation* is a theory which describes how far along the spectrum of available institutions the state will invest in order to manage the exploitation of a given natural resource. The *optimal institution* (from the owner-state's perspective) is the one that balances the gains from investment in these natural resources, against the costs of forgone alternatives.

Therefore, the fundamental reasons for instituting an open access regime, and thus indirectly generating overexploitation, are those which determine the incentives for relative investment rates for particular forms of natural resources. Relative unattractiveness for investment purposes denies institution-building investments to a wide variety of diverse natural resources, resulting in 'open-access' regimes by default, and consequential overexploitation.

This explains why 'obviously inefficient' management regimes (from a global perspective) are nevertheless implemented by the local regime. An obvious example is the case of the African elephant, where a few African owner-states implemented open access regimes, resulting in large-scale overexploitation and an international outcry. The environmental problem here is an example of the divergence between the local and the global optima, not an example of sub-optimal management. From the perspective of these owner-states, the flows of appropriable values from the elephant did not warrant large-scale investments in management. From the perspective of the international community, the flows of stock-related values (e.g. existence value) made the implemented management regimes appear obviously inefficient. It is this asymmetry in perspectives (derived from the asymmetry in appropriated values) that lies at the heart of many international conservation problems.

5.4.1 The General Model of Bio-Institutional Regulation

There are three distinct layers to the bio-institutional regulation problem. At the base are the fundamental state variables; that is, the state of the biological and institutional systems determining the bio-institutional nature of a particular commons. Interdependent individuals (i.e. those working in the context of joint assets) form the second layer of the problem. They are constrained to make their choices within the context of the current state of the bio-institutional system. Finally, in the third layer is the regulator (in this context, the owner-state). The regulator is constrained by the decision-making processes of the interdependent individuals, but at the same time has input into that decision-making process both directly (through direct subsidies to their choice variables) and indirectly (through investments in the shaping of their base institutional environment). This is the role of the natural resources regulator: to invest optimally in the institutional environment so as to develop incentives for investments in the commons' natural resources.

This can be depicted in the following model, which is the generalised version of bio-institutional regulation.

The General Model of Bio-Institutional Regulation
Regulator's Objective

$$M_s^* : \quad \underset{M(M_s)}{\text{Max}} \int_0^\infty e^{-rt} \left[\sum_{i=\mu}^n \prod_i (y_i(M), R_i(M), M_i(M); x(M), R(M), M) - r\rho_M M_S \right] dt \quad (5.15a)$$

subject to:
Interdependent Individual Choices

$$y_i^*(M), R_i^*(M), M_i^*(M) : \quad \underset{y_i, R_i, M_i}{\text{Max}} \int_0^\infty e^{-rt} \left[p(y)y_i - c(R, M)y_i - r\rho_R R_i - r\rho_M M_i \right] dt \ \forall \ i \quad (5.15b)$$

subject to:
Bio-Institutional Framework

$$i = \mu(M), ..., n \quad \text{where } 1 \leq \mu \ (M(M_s)) \leq n; \ \mu' > 0, .\mu'' < 0 \quad (5.15c)$$
$$\dot{x} = H(x, R) - y$$

At the base of this model lie institutional investments by the state (M_S), occurring either *directly* (M_S^D) via investments in monitoring and management, or *indirectly* (M_S^I) via investments in incentive mechanisms to induce individual investments (M_i). The aggregate of individual and state investments (M) is the total amount spent by the society on institution-building.

Institution-building investments have their impact by altering the bio-institutional framework within which individuals make their decisions (5.16c). Specifically, it has the impact of reducing the number of *unco-operative strategies* with the commons. This does not necessarily occur by virtue of excluding individuals from the environment (i.e. the property rights solution), only by removing non-cooperative strategies. Full co-operation occurs when a group of interdependent individuals internalise their externalities, and thus behave as if they were a single operator in the commons. This is accomplished by means of investments in institution-building in this model. Increasing investments in institution-building have the effect of increasing the number of interacting individuals in the cooperating group ($\mu = \mu (M)$; $\mu'{>}0$).

For example, cooperation within the commons may be conceived of as a group of the first $1 \ldots \mu$ (of n) individuals banding together and delegating all of their rights to one of them. This is assumed to fully internalise the externalities between members of that group. The delegate might then distribute the results of its management to the group members in accord with previously-agreed individual shares. Alternatively, the cooperating group might agree a joint management plan for the natural assets, and then distribute enforced permits for individual shares corresponding to this aggregate quota. Either avenue is capable of generating full cooperation among this group.

Increasing investments in institutions will reduce the number of non-cooperative strategies within the commons ($\mu'{>}0$), but just as in the fence-building analogy of section 5.2, it is increasingly costly to procure the cooperation of the remaining non-cooperators (i.e. $\mu''{<}0$) (on account of their relative abilities as 'non-cooperators' and on account of the increasing returns to – and increasing motivation for – non-cooperation) (Barrett, 1989; Hoel, 1990).

Thus, there will probably always remain a residual of non-cooperative strategies in the commons ($n - \mu$). This residual will include all manners of uncooperative behaviour: those who decline to join the cooperative group and those who join the group but still attempt non-cooperative behaviour. It is this residual of non-cooperative behaviour that adversely impacts the incentive structure for the decision-making by the individuals in the commons.

This effect can be seen in the levels of investment in the natural assets that result, after the level of investment in management by the owner-stage is implemented. This decision will determine the level of management spending by the individuals in the commons, and hence determine the aggregate level of management spending (M). This management spending then determines the residual amount of non-cooperation within the commons (i.e. $n - \mu$). Individuals harvesters then take their decisions on investment in the natural assets: the base resource (R), and its biomass (x). The resulting investment levels are shown in equation (5.16a and b).

Consequential Investments in Natural Resource
Stock Investments

$$x^* : \lambda \cdot \frac{\mu}{n-\mu} H'_x - \dot{\lambda} \cdot \frac{\mu}{n-\mu} = r\lambda \qquad (5.16a)$$

Base Resource Investments

$$R^* : \lambda \cdot \frac{\mu}{\mu-n} H'_R - \frac{c'_R y}{(n-\mu)} = r\rho_R \qquad (5.16b)$$

$$\text{where } \lambda = p\left[1 + \frac{1}{\varepsilon_d(n-\mu)}\right] - c$$

The implication of the equations in (5.16) is that investment in the natural assets is wholly dependent upon the level of residual non-cooperation within the commons, for two reasons. First, each individual in the group of cooperators views the returns from its investments as being distributed across the number of *non-cooperative* strategies within the commons (i.e. $n - \mu$), and there are only μ (cooperative) individuals considering investment. Hence, the returns from investment are internalised by only a factor of $(\mu/n - \mu)$. Secondly, the benefits from cost reductions are averaged over the entire next harvest's yield, so once again this return from investment is averaged over the number of non-cooperative strategies. The 'residual of non-cooperation' remaining in the commons (by virtue of the owner-state's investment decision) is an important factor in determining the aggregate level of investment in the natural assets.

The relationship between the general bio-institutional model and the special models presented earlier in this chapter lies in the generalisation of the *institution-building function* ($\mu(M)$). This function provides the link between investment spending and management systems. In essence, this function states that increased commitments of resources create enhanced management and incentive systems for enhanced management, thereby resulting in the removal of non-cooperative strategies from the commons.

For example, with investments in institution-building going to infinity, it is even possible to achieve first-best natural resource exploitation through the coordination of all activities in the commons. This may be checked by the substitution of the term ($\mu = n - 1$) into equation (5.16); when there is no residual non-cooperation, the incentives for investment in the natural assets are 'first-best'.

Achievement of First-Best Natural Resource Management

$$\text{For } M \to \infty \longrightarrow \mu \to n \longrightarrow x^*, R^* \to x^{fb}, R^{fb} \qquad (5.17)$$

The regulator will not, of course, incur infinite levels of investment in the pursuit of first-best incentives. The regulator will instead trade-off the benefits from further coordination with the costs of additional institutional investment. This is precisely the same condition derived before, in section 5.3.

Optimal State Investment in Institution Building

$$M^*(M_S^*): \int_0^\infty e^{-rt} \left[\sum_{i=1}^n \frac{\partial \prod_i}{\partial M} \right] dt = r\rho_M \qquad (5.18)$$

Therefore, the general model of bio-institutional regulation provides that the state will first determine the optimal level of investment in institutions, with individuals reacting in a predictable fashion to the revised bio-institutional environment. The result of the state's choice is a determinate institutional regime for regulating the base natural resources. The result of the interacting individuals' choices is a determinate level of investment in the natural resources. The ultimate result is a perceived level of overexploitation in the commons.

5.4.2 The Fundamental Environmental Problem of Overexploitation

This chapter commenced with a discussion that argued that the existence of an open-access regime is not a fundamental explanation for overexploitation and resource decline. The purpose of this chapter has been to address the resulting question: What are the fundamental forces driving overexploitation? The general answer has been the incentives to underinvestment.

Open-access regimes are not fundamental causes of overexploitation; they are only proximate causes. In the context of the general model of bio-institutional regulation, the nature of the regulatory regime applied to the commons is determined by the level of investment in management. If the level of institution-building investment goes to infinity, then a first-best incentive system results. If the level of such investment goes to zero, then an 'open access' incentive system results.

Investment Determined Incentive Systems

$$If \ M \to \infty \quad \longrightarrow \mu \to n \quad \longrightarrow x^*, R^* \to x^{fb}, R^{fb}$$

$$If \ M \to 0 \quad \longrightarrow \mu \to 1 \quad \longrightarrow x^*, R^* \to x^{OA}, R^{OA} \tag{5.19}$$

Of course, these two management systems are polar representations of the range of institutions available for the management of the commons. It is not always the case that such corner solutions would result from the solution of the state's investment choice problem. The aggregate level of management spending on a given resource will lie somewhere on this spectrum, and will determine the investment incentives for that resource.

However, it is nonetheless apparent that in many circumstances it is in fact the case that OA regimes (or regimes closely approximating them) are applied to many forms of natural resources. This occurred in the context of the four African states which removed a half million elephants in the 1980s, for example, and it occurs with regard to many of the other diverse resources on earth.

These systematic refusals to invest in management institutions ultimately derive from the general forces that determine the perceived investment-worthiness of various natural resources. As was mentioned above, the 'environmental problem' in this instance is not inefficient local management, but an asymmetry of perspectives on the same problem. The international community realises many stock-related values from the existence of these resources that are not appropriable by the owner-state, and the (globally) 'optimal' level of management spending from this vantage-point is determined by these benefits (and little consideration of the implied costliness). From the perspective of the owner-state, however, the level of management spending must accord with the appropriated flows from the asset, and this is an entirely distinct set of benefits and costs to be considered.

Therefore, the fundamental problem of overexploitation, in regard to terrestrial resources, probably lies in this asymmetry in perspectives on the same problem. The divergence between local and global optima regarding particular resources causes the locally-applied level of management spending to appear 'optimal' from one perspective and 'suboptimal' from the other. The regulation of overexploitation will require policies that directly address this underlying problem.

5.5 CONCLUSION

Overexploitation occurs in many different nations, but there is no state in which all forms of natural resources are overexploited. This is because resource overexploitation is simply another way of exhibiting that the resource is attracting insufficient investment. The failure to invest in a natural resource, its base and its stock, determines that it will be the subject of overexploitation.

Similarly, the failure to invest in investment-inducing incentives will determine that the resource suffers underinvestment, and hence overexploitation. It is the institution-building process that generates these incentives, and it is the investment in resource-specific institutions that determines whether particular resources are overexploited or not. The vast majority of diverse biological resources currently languish outside effective management systems for lack of such investments. The application of open-access and similar regimes for overexploitation simply reflect the owner-state's unwillingness to make those investments a high national priority (given the level of appropriable benefits from that resource and the costliness of its management). From the 'local perspective' (of that owner-state), this is the optimal level of investment for that resource.

Then what is the 'environmental problem' concerning these declining diverse resources? The environmental problem stems from the asymmetry in perspectives on the declining resources, between the local and the global communities. And this asymmetry is also easily traceable. Local communities respond to locally-received benefits from the resource (and the opportunity costs of increased investments). Global communities perceive the globally-received benefits from the resource (and probably little of the opportunity costs of increased investments).

Therefore, the environmental problem involving diverse resource overexploitation and consequent decline in many states is an international one. It concerns the necessity of creating mechanisms for registering the preferences and concerns of the global community with the local. This is the path for regulating overexploitation.

6 The Global Biodiversity Problem:Optimal Policy and the Global Conversion Process

6.1 INTRODUCTION

The problem of global biodiversity losses derives from the failure of states to consider the impacts upon global stocks when taking national regulation decisions. When stock depletion generates substantial externalities, as is the case with diverse biological resources, then decentralised (i.e. multinational) regulation of these global stocks will necessarily be inefficient. The biodiversity problem is, in effect, the predictable result of an imperfection in the existing global regulatory system with regard to diverse resources; decentralised decisions regarding diverse resources cannot take into account the innate value of their diversity.

The statement of the solution to the global biodiversity problem is straightforward: decision-making in regard to diverse resources must take these stock externalities into consideration. However, this solution concept runs into the same difficulties as the underlying biodiversity problem, because it also must proceed through the same decentralised regulatory system. As natural resources are *national* resources, the mere recognition of these regulatory imperfections goes little further toward their solution. It is instead necessary to construct global incentive mechanisms which induce national regulatory actions consistent with the global object.

This chapter outlines the general nature of the global incentive mechanism required to internalise the value of biodiversity, but only in section 6.5. First, it is necessary to provide a clear depiction of the nature of the problem to be solved; this description will then suggest the nature of its own solution. The first three sections of this chapter undertake the task of formulating the nature of the regulatory problem that generates inefficient global biodiversity losses. In section 6.2, the human-induced extinction process is described as the result of human choice in the conversion of the biosphere. In section 6.3, the predictable failures in the decentralised regulation of this conversion process are described. In section 6.4, the costliness of these failures (i.e. the value of biodiversity) is outlined. Then, finally, in section 6.5, the general nature of the international incentive mechanisms necessary to redress these regulatory failures is discussed.

The international regulation of extinction may sound like an unlikely or at least impractical project. However, the point of this chapter is that the regulation of extinction is occurring right now, only on a *national* basis and with a demonstrably suboptimal outcome. The international community needs to intervene, in certain specific ways, in order to redress the imperfections in the existing regulatory system.

6.2 REGULATING THE GLOBAL BIODIVERSITY PROBLEM: PAST, PRESENT AND FUTURE

For regulatory purposes, terrestrial natural resources are first and foremost national resources. This is the meaning of *the doctrine of national sovereignty* in international law. Each state has the unrestricted and exclusive right to determine (through act or omission) the management of the various natural resources that it 'hosts'. There is good reason in economics, as well as in law, for such a doctrine; it serves the role at the international level that property-right institutions serve at the domestic. That is, under the doctrine of national sovereignty, there is a specific state that has the designated responsibility for the management of any given area of terrestrial resources. As with property rights, this is a useful mechanism for allocating management responsibilities to those agents best able to invest optimally in the resources.

However, just as with domestic property rights, problems can arise where environmental systems extend beyond institutional boundaries. This is the nature of the problem of global biodiversity losses. It is a problem resulting from the decentralised regulation of a process with clear global implications. In essence, the problem derives from the fact that the impacts of *national* resource exploitation on *global* stocks are not considered by individual regulator states. Therefore, the general problem of global biodiversity losses concerns the difficulty of adequately regulating global stocks of diverse resources in a decentralised (i.e. multinational) world.

Before moving to the issues of optimal international regulation, it is necessary to understand the nature of, and imperfections within, the existing, decentralised regulatory process. This section presents a specification of the existing global conversion process, its decentralised stopping-point, and the reasons for its divergence from the global optimum. This constitutes a specific representation of the *global biodiversity problem*; it is a problem of a global process that is inadequately regulated in a decentralised (i.e. multinational) world.

6.2.1   The History of Biosphere Conversion: Inefficient Decentralised Regulation of the Global Conversion Process

This section tells a very short story about a very long period of time. Its purpose is to develop a schematic that is descriptive of the process by which the earth's biological resources have been converted. It encapsulates a 'history' of the biosphere since the introduction of domesticated and cultivated species, some ten thousand years ago.

The history of the biosphere since the time may be best described as one of *sequential conversions: the alteration of the world's slate of natural resources on a state-by-state basis.* Conversion within a particular state has occurred as and when the idea of employing domesticated and culti- vated varieties (and the methods of production that they imply) has arrived at that state at a particular point in time; then, the natural slate of resources is usually displaced by the domesticated. The sequential nature of this pro- cess is discernible in the non-uniform progress of the process across the world's states; some states are long-converted, while others have had little resource conversion within their borders. The conversion process has therefore diffused in general much more rapidly within particular states than between them. In order to simplify the exposition in this chapter, it will be assumed that the conversion frontier moves between states, rather than within them, and that (on arrival of the frontier) the adoption of the ideas of modern agriculture (i.e. conversion to specialised varieties) is of the nature of an 'all or nothing' proposition; this is the 'sequential con- version process' analysed here.

Therefore, in this analysis, the time horizon relating to the biosphere is demarcated into two sections, representing the succession of states that have already converted their biological resources (the 'past') and those states which have not (the 'future'). The conversion frontier (the 'present') resides at the *marginal state (MS)*: the state that is currently contemplating the conversion of its natural slate of biological resources.

The force that generates the progression through this series of states is the *global conversion process*. It is driven by the perceived net benefit from the conversion of the state's naturally-evolved slate of resources to the human-selected specialised slate (i.e. the slate of domesticated and cultivated varieties and their associated methods of production).

Over the course of the history of the biosphere, each state presented with the idea to date has undertaken the option of conversion, i.e. the net benefits to conversion have been perceived to be positive. With successive conversions, it is likely that the perceived net benefits to conversions will decline, on account of both rising supply costliness as well as falling prices for the goods and services flowing from conversion. However, with so few states remaining within the non-converted margin, it is apparent that the perceived net supply costs of conversions are not rising rapidly enough to deter conversions by themselves.

This process is depicted in Figure 6.1. Along the horizontal axis is the passage of time, representing sequential conversions of the diverse resources managed by individual states. The history of the evolved biosphere ends at the point of 100 per cent conversion, when no conversion frontier remains on earth. The question addressed in Figure 6.1 is whether the global conversion process will continue through to totality (the elimination of the evolved biosphere), or whether there are forces that will halt this process.

Figure 6.1 depicts a scenario where there is little likelihood that the conversion frontier will be halted. First, it assumes that each state per-

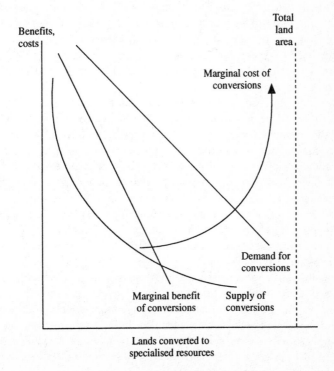

Figure 6.1 Optimal policy regarding conversioins

ceives a non-increasing costliness to the conversion of its natural resources; that is, the costs of the specialised inputs (machinery, chemicals and species) of converted production are non-increasing to each successive state. This is represented by the downward-sloping *perceived supply curve* in Figure 6.1, which is the perceived marginal cost of conversion at each point in time (i.e. by each state).

Another reason why there is little reason that conversion will halt is the lack of a demand-related constraint on the process. The analysis within Figure 6.1 assumes that there is no constraint arising from a lack of subsitutibility between specific natural and specialised outputs, although the average benefit from additional units of identical outputs is declining. This is represented by a downward-sloping linear *demand curve* (representing the perceived average benefit from conversion to specialised resources). The average benefit to conversion declines, because of the existence of some diverse products for which consumers hesitate to accept specialised substitutes. Then, as global conversion approaches totality, the marginal benefit to further conversions becomes very small due to consumer hesitancy to substitute, but never small enough (in Figure 6.1) to halt the process.

It is important to be precise concerning the nature of the 'benefits to conversion', as they are only in the short run derived from the increased productivity of specialised resources. In the medium run (with other factors adjusting), the benefits from conversion are derived from very different sources.

In the medium run, it is almost certain that the demand curve for conversions would be downward-sloping. The first reason for this is derived from the argument for conservation advanced by John Krutilla (1967). He argued that as human development continues, by reason of the conversion of natural resources, one of the few remaining scarce resources will be 'naturalness' itself. In short, the only value of the increased wealth from conversion lies in the increased quantities of goods and services that may be consumed, and Krutilla hypothesised that there is a trade-off inherent in development between quantity and quality. Therefore, increasing wealth, with increasing conversion, itself may imply a decreasing relative value for the products it can generate.

The alternative to increased wealth (from increased conversions) is increased fitness. If the benefits from conversion were utilised for this purpose, then the primary effect of the conversion process would be to expand the human niche. This means that, in the longer run, the benefits flowing from land conversions depend entirely upon the social benefits to be derived from greater population densities. Does a society benefit more from higher or lower population densities?

There is, of course, a long debate concerning the benefits to be derived from population growth. One side to the debate points to the historical evidence that population growth has been the catalyst of economic and social change over the past few centuries, with significant societal benefits (Simon, 1992). The other side to the debate looks at the present and to the future, and finds little prospect for societal benefits in further population growth (Lee, 1991).

It seems possible to explain the difference in perspectives that generates both sides to this debate. In short, the benefits to higher population densities probably depend on the population densities that exist in the other societies on earth. For example, if the average benefit curve to further conversions was assumed to be downward-sloping in the longer run (with populations adjusting), this implies that the strategy of high population density (for a given state) would initially afford substantial average benefits (in an otherwise low-density world). This would be the case if the benefits were received by reason of the *relative uniqueness* of the capabilities (urbanisation, industrialisation) that high density would afford over low-density societies.

However, if relative uniqueness is the underlying rationale for population-based benefits, then, as additional states convert to the same strategy, the average benefits would decline. This is because the new high-density populations would be ready substitutes for one another in the population-based services (e.g. low-cost labour-based manufacturing) that they can provide. In addition, the *marginal benefits* of a high density strategy would decline even more rapidly than would the average, and these are the source of a pecuniary externality operating between the newly-converted and the formerly-converted states.

Therefore, in this framework, there is a germ of truth to both sides to the debate. Looking retrospectively, societies operating with high population densities (e.g. Europe and Japan) in a low-density world have perhaps gained substantial advantages from the distinct opportunities that this factor has provided. However, with a large part of the world already converted, the average benefits from further conversions must now be much reduced. Looking solely prospectively (from the vantage point of the marginal state forward), there is a much reduced benefit to individual land use conversions and the higher population densities they imply, even though in the past the benefits were much greater. The position that one takes in the debate simply depends on where within the global conversion process one stands when considering the average benefits to conversion (and hence high density).

This might indicate that the 'population pessimists' have the upper hand in the debate within this model, since they seem to be focusing upon the more relevant notion of 'average benefit', i.e. that average benefit which exists in the present and future. However, it is more likely that the pessimists are looking to the marginal than the average benefit to population growth, and deriving their conclusions from this. This marginal benefit to population growth (which is perceived by the converted but not the unconverted states) will fall much below the average benefit. This is attributable to a pecuniary externality, which drives the value of the uniform specialised outputs downwards with each successive entrant. A difference in perspective here (in terms of focusing on marginal versus average benefits) might explain the ironic nature of the current debate, with the pessimists within the relatively high-density states (of e.g. Europe) advocating the benefits of population control to the relatively low-density states (of Africa and South America). One group might be pointing to the marginal benefits to such population expansions, while the other is reacting to the average.

In any event, all of these considerations indicate that the average benefit curve for further conversions (and hence population expansions) is downward-sloping. However, if the demand for the flows from specialised species is not too inelastic (indicating that there is broad substitutibility between species for basic needs such as food and clothing) and increasing returns to scale are perceived to apply to a handful of species over the entire global land area, it is possible that these conversions might continue through to totality, as depicted in Figure 6.1. Then the entirety of the output from the natural evolutionary process will be lost to the global conversion process.

6.2.2   The Future of Biosphere Regulation: International Institution-Building for Internalising Externalities

There is another force which should enter into decision-making concerning the conversion process. This is the geometrically-increasing costliness of the final conversions of diverse resource stocks (or, alternatively, the marginal value of biologically diverse resources). This is represented by the upward sloping *actual supply curve* in Figure 6.1, which includes the marginal opportunity cost of the loss of these diverse stocks of lands and resources.

This is an important externality within the decentralised regulatory process. At each point in time (i.e. with each successive conversion), this costliness is increasing but unaccounted for, because the opportunity costs of conversions in terms of lost global services are not included within the converting state's decision-making framework. The 'marginal state' considers only the perceived supply cost, not the actual cost of conversion.

The loss of diverse resource stocks necessarily entails social costliness, in terms of irreplaceable insurance and information services, and hence these losses should be considered as an opportunity cost in the supply of converted lands. The precise nature of these costs is described in section 6.4, below. In fact, the marginal costliness of successive state conversions will rapidly increase, once a substantial part of the terrestrial surface has been converted to specialised resources. This is because the relationship between conversions of land area and loss of species stocks is non-linear. Species-area functions are said to follow an Arrhenius (log-linear) relationship. Studies in island biogeography demonstrate just such an empirical relationship, indicating that the conversion of 90 per cent of land area results in losses of about 50 per cent of the species diversity. The importance of this relationship lies in its implications regarding the final conversions of the residual territory.

In general, this final 10 per cent of conversions will entail as great losses of diverse resources as did the initial 90 per cent. Therefore, to the extent that global costliness is directly related to the loss of species diversity (as will be shown to be the case in section 6.4), this costliness will increase geometrically with the final conversions.

With the passage of time (and successive conversions), there are two countervailing forces which should determine the globally optimal stock of biodiversity: the benefits from specialisation and the benefits from diverse resources. The global conversion process should halt when the marginal global value of the next conversion is negative, or alternatively, when the marginal value of retaining the resources in an unconverted state is positive.

In the context of Figure 6.1, the global process of conversion should be halted by the force of the value of global biodiversity at least at the intersection of the actual supply curve with the demand curve (at time t* and state MS*), but it is not on account of the decentralised nature of the process. An optimal international policy for biodiversity would then attempt to target this point through regulation of the global conversion process.

An optimal international policy for biodiversity must halt the global conversion process at the optimal point in time (at the optimal 'marginal state'), and it must do so (in a multinational world) through influencing the incentives perceived by those states. International institutions for biodiversity conservation should have as their objective the substitution of the 'actual' for the current 'perceived' cost curve within the decision-making of marginal states. That is, international institution-building must take the form of investments in re-channelling the global values of diverse resources to the host states themselves. In this way, it is possible to achieve an optimal amount of conversion within a multinational world.

6.3　OPTIMAL POLICY AND THE GLOBAL BIODIVERSITY PROBLEM

This section makes concrete the various concepts developed within section 6.2, in order to define the nature of an optimal biodiversity policy. Given the dynamics of the global conversion process (as described in section 6.2), the problem may be defined as the choice of an optimal 'stopping rule' with regard to this process. That is, optimal policy for biodiversity devolves in its simplest form to a determination of the *marginal state* at which the global conversion process is optimally halted. This section gives a model for this optimal policy, and describes the nature of the intervention required to implement that policy.

6.3.1　Optimal Policy for Biological Diversity – A Model

As developed in section 6.2, the global biodiversity problem is best-described as a conflict between nationally rational and globally optimal decision-making with regard to the global conversion process. This process may be conceived of as a *sequential bio-economic process*. The passage of time in this model captures three important dimensions within this process: the movement of the 'conversion frontier' to the borders of the next state; the change in the value of the remaining diverse resources (as global stocks erode); and the accumulation of information. The conflict arises on account of the predictable divergence between the local and global optima. The local 'host state' will make a decision regarding the conversion of its diverse resources which does not consider the global benefits derived from these stocks, and this externality is the source of the global biodiversity problem.

Specifically, imagine that the states of the earth form two groups at any point in time, termed 'North' and 'South', the former being those which have previously converted their lands to specialised biological resources and the latter being those which have not. The issue (for purposes of optimal policy) is whether the marginal state (MS) in the group South should convert its biological resources when full global costliness is considered. The globally optimal stopping rule will halt the global conversion process at the point in time (i.e. the marginal state) where the marginal benefits are in equilibrium, and irrespective of the local optimum.

In order to depict the optimal policy applicable to this process, imagine that, as the idea of biological specialisation diffuses across the earth, each individual state (s) determines whether to convert its naturally existing vector of biological resources (x_s) to a specialised slate of resources.

Conversion consists of eliminating the land area dedicated to non-specialised biological resources (R_s), and coincidentally the state's stocks (x_s) of diverse resources are also replaced by specialised stocks. The reduction in global stocks of diverse habitats (R) and resources (x) has impacts both on the individual converting state (primarily in terms of changes in appropriable flows), and also on the welfare of all other states (primarily in terms of non-appropriable insurance and informational services).

The driving force behind such stock conversions is the nature of the flow generated by the different biological stocks. Specialised biological resources generate flows to their host state (y_s^{sp}), a vector of flows corresponding to a subset of species taken from a small menu available to all states (Y^{sp}). The important point about this vector is that it captures nearly the entirety of the flows emanating from that state's specialised stocks (x_s^{sp}). The degree of local appropriability of these flows is virtually 100 per cent.

On the other hand, a diverse stock of resources generates a vector of flows to the host state (y_s^d) which is very different from the specialised vector available through conversion. First, the dimensionality (in terms of species) of the individual state's 'diverse resources vector' is several orders of magnitude greater than that of the 'specialised resources vector' (thousands of species as compared to dozens). Second, the rate of appropriability is far greater than it is for the diverse resource vector; that is, the appropriated flows (y_s^d) from diverse resource stocks (x_s^d) represent only a small proportion of the total flows from these stocks.

In addition, the global stock of diverse resources (x) generates a vector of flows (Y^d) which is again several orders of magnitude greater than the dimensionality of y_s^d in any one state; that is, there is little intersection between the diverse resource slates in the various unconverted states. In contrast, the global dimensionality of the menu of specialised species (Y^{sp}) is only in the dozens, and the Northern states have substantial commonality in flows.

The global conversion process can then be conceptualised as the sequential decision-making by successive 'marginal states' on whether to 'change menus', from diverse to specialised resources. If conversion is elected, this is accomplished by means of non-investment in diverse resource stocks (resulting in their removal by overexploitation and replacement), and contemporaneous investment in the specialised species (by selecting some subset from the global menu, Y^{sp}).

Diverse and Specialised Resources – Definitions (6.1)

$R_{(Sx1)}$ ≡ the area of diverse habitat in each of the S states in group South

$x_{(SxD)}$ ≡ the stocks of diverse resources in each of the S states in group South (D diverse species globally)

$Y^{sp}_{(NxC)}$ ≡ the aggregate flows from global stocks of specialised resources (C specialised species globally)

$Y^d_{(SxD)}$ ≡ the aggregate flows from global stocks of diverse resources

$y^{sp}_{s\,(1xc)}$ ≡ the appropriable flows from each of the c specialised species within (converted) state

$y^d_{s\,(1xd)}$ ≡ the appropriable flows from each of the d diverse species within (unconverted) state

The optimal policy for biodiversity is the determination of the optimal stopping point in the global conversion process. The problem is to select the marginal state (MS) at which the conversion process should be halted, if the maximisation of aggregate global benefits is the objective.

The Regulation of Global Biodiversity – Optimal Conversions

$$MS^\Omega : \frac{\text{Max}}{MS} E\left[\prod_G^{MS}\right] \equiv$$

$$E\left[\sum_{s=1}^{(MS-1)} \prod_s (y^{sp}_s; x^{MS}, R^{MS})\right] + E\left[\sum_{s=MS}^{N+S} \prod_s (y^d_s; x^{MS}, R^{MS})\right] \qquad (6.2)$$

where MS ≡ the marginal state member of group South

$\prod_s$ ≡ the present value of the flow of benefits to state s

x^{MS}, R^{MS} ≡ the global stocks of diverse resources and habitats given marginal state (MS)

$N+S$ ≡ the number of sovereign states (decentralised management units)

$\prod_G^{MS}$ ≡ the present value of the global benefits with marginal unconverted state MS

The two terms in equation (6.2) represent the total benefits to the group North and the group South, respectively. The act of conversion shifts the marginal state in group South (MS) to the group North (implying that the second term then commences summation from $S = (MS + 1)$). This means that the marginal state has converted its biological assets to some subset of the specialised roster of resources.

This has the nationally perceived effect of shifting production from diverse to specialised appropriable flows (i.e. from y^d_s to y^{sp}_s). It also has a global impact, in terms of a reduction in the global stocks of diverse resources (from x^{MS}, R^{MS} to x^{MS+1}, R^{MS+1}). Since global stocks of diverse resources contribute to production in both sectors, while only a portion of this flow is locally appropriable (in the form of y_s^d), the globally optimal stopping rule in the conversion process must take into account this positive stock externality to the Northern states.

Therefore, globally optimal regulation of biological diversity takes the form of halting land use conversions, even when these are perceived to be in the national interest, if they are contrary to the global interest. Because the conversion process has occurred historically on a state-by-state basis (rather than uniformly across all states), the globally optimal stopping rule also has the effect of determining the final size and constituencies of the groups 'North' and 'South'. That is, it creates a 'division of labour' in world biological production, with one group of states specialising in the production of diverse resource flows and the other specialising in the production of specialised resources.

6.3.2 The Nature of Optimal Intervention

It is important to be precise about the meaning of the vector y_s^d – the flow of goods and services which come within the value function of the host state – as it is this definition which captures the essence of the necessary intervention. The vector y_s^d *is best conceptualised as that part of the flow from diverse stocks which is appropriable by the host state under existing institutions.*

Some of the benefits of diverse resources are flows, but flows whose appropriation by the host state (via unilateral action) is comparatively costly (or impossible). To the extent that these non-appropriable flows constitute a significant portion of the benefits from diverse resource stocks, there is an inefficient bias against holding these forms of biological assets.

As states continue to change the form of their holdings of biological assets, always choosing from the same small slate of specialised resources, the biosphere on earth assumes a very particular form. There are increasing numbers of a very small subset of species on the earth (as the group North grows) and there are decreasing numbers of all other species (as the group South shrinks). Given the role of diverse resources (shown in the next section), this implies that the non-appropriable values of diversity will be increasing with successive conversions.

This determines the nature of the intervention necessary for the optimal regulation of extinction. The objective is to install an optimal stopping rule to the global conversion process, so that this process is halted when the marginal values of diverse and specialised resources are in balance. One mechanism which will implement such a solution is to bring more of the non-appropriable stock-related values within the host state's decision-making process, i.e. to internalise the global externalities of diverse resource stocks. In terms of the model outlined above, this is accomplished by means of shifting more of the flow from diverse resource stocks (x_s) into the appropriable flows vector (y_s^d). The impact of internalisation will be to cause marginal states to perceive the level of opportunity costs implicit within the marginal cost curve in Figure 6.1, inducing globally optimal decision-making by individual states. This has the effect of implementing a stopping rule at the globally optimal point in the conversion process.

This is the probable nature of the required solution to the global bio-diversity problem for two reasons: first, since diverse resources all lie within sovereign states, the mode of intervention must operate through the decision-making frameworks of the host states; secondly, so long as all states have a right to development, this solution concept protects that right in a way consistent with the protection of global values. In short, focusing upon enhancing the appropriability of global externalities affords these states an alternative path to development, rather than denying 'Southern' states development in order to maintain the global optimum.

Therefore, the general nature of the global biodiversity problem is the presence of significant levels of non-appropriable benefits from the main-tenance of diverse resource stocks. The general nature of the solution is the internalisation of these same benefits within the decision-making process of the marginal states.

Clearly, the existence of such stock-related (i.e. non-appropriable) flows is a necessary condition for the existence of either a problem of biodivers-ity, or a solution. The existence of these benefits, and their nature, is the subject of the next section. In section 6.5, the chapter returns to the subject of the optimal form of international intervention.

6.4   THE VALUES OF GLOBAL BIOLOGICAL DIVERSITY

The value of diverse biological resources is, in terms of the model of the previous section, equal to the total flows from all of the diverse resource stocks that exist on earth (x). This is the value that should be given effect within the context of an optimal regulatory policy, in order to halt the global conversion process at the globally optimal point. This value is defined in the next section as the 'marginal value of biological diversity (MVBD)'. It is a functional notion of value alone, devised because it is required for the purpose of defining the regulatory objective. This concept is the subject of the first part of this section.

Before proceeding to a discussion of this concept, it is important to note the distinction between the values of biodiversity and diverse biological resources, as the two are not synonymous. The latter is an all-inclusive category, encompassing the tangible and intangible flows from all biological resources that exist in areas that have not been subject to human conversion. In contrast, the latter term corresponds only to the value of 'diversity' (as opposed to 'uniformity' or 'homogeneity'); it is the value that flows from the mere fact of non-conversion.

In this sense *biodiversity* contains two distinct components of value: first, there is the unique value of the goods and services generated by the evolutionary process (as they encapsulate a 4.5-billion-year history of adaptation and coevolution); and second, there is the value of retaining a production strategy on earth distinct from the 'specialised' one. These values, i.e. the values of global biodiversity, are the subjects of the latter three subsections of section 6.4.

6.4.1   The Marginal Value of Biodiversity (MVBD)

A concrete meaning may be given to the concept of *the total value of global biological diversity*. This is the opportunity cost of the conversion of diverse resources to specialised ones. There are both 'total' and 'marginal' concepts to be distinguished. The total value of global biological diversity would correspond to the area under the actual supply cost curve in Figure 6.1. *The marginal value of biological diversity* (MVBD) would correspond to a specific point on this curve, i.e. the costliness of taking the marginal step in the conversion process.

In both cases the concept corresponds to the opportunity cost of further conversions. However, the distinction is important because one is an operational concept and the other is not. It will never be possible to give precise meaning to the concept of the total economic value of biological diversity. This is because the diversity of life forms on earth is one of the fundamental components for maintaining stability in the biological production system sustaining human societies. As diversity goes to zero (total global conversion), the level of instability introduced implies that there is little prospect for human production systems to sustain themselves over any significant time horizon. Therefore, as conversion spreads to the last corners of the earth, its opportunity cost must be unbounded, as all human-sourced values must depend upon the maintenance of the biological production systems that sustain human societies. This is represented in Figure 6.1 by the area under the actual supply cost curve, which is unbounded as conversions approach totality.

The functional notion of economic value as applied to biological diversity is the marginal one. The value of biological diversity cannot be divorced from the sequential decision-making process of which it is an essential part. MVBD is the global opportunity cost of the marginal state switching from group South (the unconverted suppliers of diversity services) to group North (the previously converted states). It is also the force which must be effected if the global conversion process is to be brought to a conclusion short of total conversion.

The Marginal Value of Biological Diversity (MVBD)

$$MVBD \equiv \prod_G^{MS} - \prod_G^{MS+1} \tag{6.3}$$

$$given : \prod_G^{MS} \equiv \sum_{s=1}^{(MS-1)} \prod_s (y_s^{sp}; x^{MS}, R^{MS})$$

$$+ \sum_{s=MS}^{N+S} \prod_s (y_s^d; x^{MS}, R^{MS})$$

where $\prod_G^{MS}$ $\equiv$ the present value of global benefits with marginal state *MS*

x^{MS}, R^{MS} $\equiv$ the global stocks of diverse resources (lands) with marginal state *MS*

$N+S$ $\equiv$ the number of states regulating terrestrial production

Equation (6.3) states that the MVBD, at a given point in the global conversion process, is the difference between the global present value of all biological production given existing stocks of diverse resources and the global present value of all biological production given the reduction in such stocks occasioned by the transfer of the marginal state between groups, i.e. the conversion of the marginal state's diverse resources.

This concept further demonstrates the nature of the global problem. It is a problem within a process, whereby continuing conversions occur not to increase aggregate overall value, but rather to increase aggregate appropriable value. There is little interest in domestic investment in supporting global values that are not channelled through the supplier state. Therefore, another means of representing the MVBD is in regard to its appropriable and non-appropriable components. Equation 6.4 restates 6.3, but it now gives the change in the flow of values in two parts: those that are appropriable by the host state (6.4a) and those that are flowing generally to the global community (6.4b).

Appropriable and Non-appropriable Components of MVBD

$$\frac{\Delta \prod_G}{\Delta y_s^d} - \frac{\Delta \prod_G}{\Delta y_s^{sp}} \tag{6.4a}$$

$$MVBD \equiv \qquad +$$

$$\frac{\Delta \prod_G}{\Delta x_s} + \frac{\Delta \prod_G}{\Delta R_s} \tag{6.4b}$$

Equation (6.4) gives another general rendition of the global biodiversity problem; that is, it restates the reason why the local and global optima diverge. This divergence is attributable to the fact that the global conversion process is currently regulated by the force encapsulated in (6.4a), while the globally optimal regulation would incorporate the sum of the two terms in that equation.

This is because (6.4a) represents the appropriable component of MVBD; it is the part of the change in flows of value that is channelled through the host state under existing institutions. This part of MVBD is generally negative, as states continue to convert their resources for the greater local productivity gains achievable through conversion to specialised resources. The potential force for halting the conversion process derives from the values encapsulated within term (6.4b) above: the non-appropriable flows from the diverse resource stocks. The value of this term is demonstrably positive, and when it comes to outweigh the value of the appropriable elements (6.4a), the process should be halted (from the global perspective). However, given non-appropriability of these values and decentralised decision-making, there is a divergence between the local and global optima.

The remainder of this section discusses the nature of the 'stock-related' (i.e. *non-appropriable*) benefits that are encapsulated within (6.4b). These are the true 'global values of biological diversity', i.e. the values that flow uniformly to the global community from the mere fact of non-conversion. In the ensuing discussion, the stock-related values of biological diversity (6.4b above) are broken down into three distinct components corresponding to their static *(portfolio) value* and also to their dynamic value (i.e. *the expected value of information in the context of retained options*). In each case the value is developed in the context of the sequential decision-making model discussed above. That is, it is the non-appropriable element of MVBD (i.e. the global value of the marginal state's decision not to convert its resources) that is being analysed.

6.4.2   The Portfolio Effect – the Static Value of Biodiversity

Once biological diversity is considered outside a deterministic framework, there is one obvious advantage which diverse resources have over specialised, i.e. their 'pooling' capacity. If the global community is concerned not only with the mean yield of its biological resources but also with its variability, then their capacity to reduce global variability (via the pooling of distinct assets) is a desirable trait.

Global conversion upon a small slate of specialised species necessarily increases the variability in global yields. This is because the aggregate variability of all biological asset yields is not the simple summation of the individual variabilities of these assets. Aggregate variability instead depends crucially upon the independence of asset yields, i.e. the absence of a systematic correlation between them. Even if each of the different forms of biological assets has the same innate periodic variability (σ^2), the aggregate variability of these assets is equal to that variance divided by the number of independent assets (e.g. σ^2/C). This is known as the *portfolio effect*, and it derives from the fact that independent variabilities will have a cancelling-out effect within the portfolio.

Consider again the sequential decision-making model introduced earlier. In this model the marginal state must choose between biological production methods with a diverse roster of outputs (y_s^d) and specialised biological production with a small slate of outputs (y_s^{sp}). Here it will be assumed, for simplicity of exposition, that all members of the group North produce the same number of specialised resources (of dimensionality c). Members of the group South each produce a much wider range of outputs (each of dimensionality d) chosen from a much wider range of possibilities (of aggregate dimensionality D). The key difference is the dimensionality of the two vectors. In matter of fact, states in group North choose their individual production vector from an aggregate vector Y^{sp}, whose dimensionality (C) is measured in dozens. The important difference is that the dimension d is probably in the hundreds or thousands, and the dimension D is certainly in the millions, in contrast to the dimensions c and C which are only in the dozens at most.

The difference in dimensionality is important for the creation of a 'portfolio effect' both locally and globally. Any state that retains its diverse resource vector rather than converting to the specialised maintains a much broader range of assets upon which to rely (d rather than c). Equally, this state is also investing in maintaining the existing dimensionality of the aggregate diverse resource vector (i.e. D). This contributes to a portfolio effect at the global level.

This may be demonstrated in the context of the framework developed above. First, it is important to note that each of the components of the production vectors (Y^d) and (Y^{sp}) is, at base, the same commodity: the flow of biomass from stocks of base resources (land). This flow might be measured in a common unit, (e.g. usable mass, volume or energy), and then these vectors would simply represent the usable flows of energy-matter derived from various forms of production. In this schematic, a 'species' is simply a distinct 'method of production' for a common production unit, i.e. usable biomass.

In fact, this is a biologically sound description of the various components of these vectors; a species does represent a distinct method (or life form) through which the sum's energy is channelled through to human societies. The 'portfolio effect' value to retaining a diversity of species thus derives from the maintenance of a range of different methods for channelling the sun's energy to human society; alternatively, the cost of successive conversions is the increasing reliance of an ever-larger human society (built upon the increased average productivity of specialisation) upon a decreasing number of available methods of production.

To demonstrate this effect and the costliness it implies, assume that the annual production of usable biomass/energy is achieved by means of the use of the range of existing species. Global product is determined by the sum of the global yields for each of the components of Y^d and Y^{sp}. That is, these aggregate production vectors may now be considered to represent D and C observations respectively on a common unit of production, i.e. the aggregate biomass/energy output from $D + C$ distinct methods of production. Given this definition, the annual productivities of all species will be a random variable (e.g. defined in terms of energy yield per unit of base resource), which may be described by reference to the various moments of its distribution in a given year.

Also, assume that there is some 'global welfare function', $W(Y^d + Y^{sp})$, which captures the global community's willingness to trade-off between mean biological productivity and higher moments. That is, it will be assumed that $W(Y)$ is a well-defined function over at least the first two moments of Y^d and Y^{sp}. Then this function merely expresses the notion that the global community is unwilling to pursue higher mean yields blindly, i.e. there must be some consideration of the higher moments of the expected flow of benefits. It is then possible to capture the nature of this trade-off, between mean production and its variability, in the following expression of the expected marginal value of biological diversity (via application of Taylor's expansion around the mean production levels).

Marginal Value of Biodiversity – Impact of Conversion on Mean and Variability

$$E[MVBD] \equiv E\left[\prod_{G}^{MS}\right] - E\left[\prod_{G}^{MS+1}\right] \cong \left[W(\bar{y}_s^d) - W(\bar{y}_s^{sp})\right] + \frac{W''}{2}$$

$$\left[\frac{\sigma_{y^d}^2}{d} + \text{cov}(Y^d, y^d) + \text{cov}(Y^{sp}, y^d) - \text{cov}(Y^d, y^{sp}) - \text{cov}(Y^{sp}, y^s) - \frac{\sigma_{y^{sp}}^2}{c}\right] \quad (6.5)$$

where: $d\ (or\ c)$ ≡ number of diverse (or specialised) species in the marginal state

$W(\cdot)$ ≡ the 'global welfare function'

$\sigma_{y^{sp}}^2$ ≡ the variability of biological production in the marginal state with specialised species

$\sigma_{y^d}^2$ ≡ the variability of biological production in the marginal state with diverse species

The first terms of equation (6.5) (i.e. the portion within the first square brackets) corresponds to the appropriable flows from conversion (as in term 6.4A). This once again captures the nature of the force for conversion, i.e. the higher expected mean yields from conversions to specialised biological assets. It is because this part of the expected value of diverse resources is negative that marginal states continue to undertake conversions.

The second term in equation (6.5) (i.e. the portion within the second square brackets) captures the global impact of the marginal conversion in terms of global yield variability; the source of the portfolio effect. This impact on MVBD is invariably positive, i.e. marginal conversions increase global variability and thus reduce global welfare.

This second part of equation (6.5) may be broken down into three distinct parts. There are three distinct forces contributing to reduced variability from the retention of diverse biological resources. First, there is a species-specific portfolio effect from the use of a nonspecialised species. A specialised species is human-selected for certain traits, and then mass-produced; much that exists in all future generations derive from this single selection, eliminating much internal genetic variability. A diverse biological resource is itself a wider portfolio for production. Second, there is a nationwide portfolio effect from retaining diverse resource. That is, domestic variability of production is reduced because the state retains a larger number of independent methods of production. Thirdly, there is a distinct international portfolio effect. The retention of any single state's diverse resources has the potential to reduce the aggregate covariance between national products.

In sum, the marginal value of biological diversity with regard to the value of diverse resources in reducing global biological yield variability is necessarily positive. The conditions that determine this are as follows.

MVBD – Portfolio Effect (PE)

$$E[MVBD]_{PE} \cong \frac{W''}{d}\left(\frac{\sigma^2_{y^d}}{d} + cov\,(Y^d, y^d) - cov(Y^{sp}, y^{sp}) - \frac{\sigma^2_{y^{sp}}}{c}\right) > 0 \quad (6.6)$$

since

(A)	$W'' < 0$	(with risk aversion)
(B)	$\sigma^2_{y^{sp}} > \sigma^2_{y^d}$	(species-specific PE)
(C)	$d > c$	(intra-national PE)
(D)	$cov\,(y^{sp}, Y^{sp}) > cov\,(y^d, Y^d)$	(inter-national PE)

These terms give precise meaning to the various components of the global portfolio effect, as described above.

First, term (6.6B) represents the species-specific portfolio effect, in that a diverse species contains a wider range of genetic variability than does a specialised. For each diverse species, the inherent variability of production is reduced by reason of this innate portfolio of distinct characteristics.

Term (6.6C) captures the idea of greater variability stemming from the use of a smaller slate of resources within the marginal state – the intra-national portfolio effect. Assuming that each species represents an 'independent' method of production, in that the production from each is not correlated with that of other distinct species within that state, a wider menu of resources will reduce intra-state variability.

Term (6.6D) captures the nature of the interaction between production methods in use in the various states. In group North, the aggregate variability of the group must necessarily be greater when productive assets are more closely correlated, because there is less opportunity for production effects to 'net out'. This implies that the covariance between member states in group North will be higher than the covariance between member states in group South. Therefore, the marginal state's decision to switch between the two groups will necessarily increase the covariance of global yields – a reduction in the international portfolio effect.

Therefore, the movement of the marginal state from group South to group North represents an unambiguous increase in global yield variability, precisely because this conversion represents the movement on a global basis towards universally more specialised production. To the extent that global variability matters to human societies, then the increasing first moment of biological yields must be set off against the also-increasing second moment of these same yields. It is important to emphasise that this is not a form of variability that can be resolved through other forms of insurance contracting, as this is increasing *global variability*. Reductions in the portfolio of diverse biological assets clearly have this effect on a global basis.

6.4.3   Dynamics: The Value of Exogenous Information

The use of this sequential decision-making model of the global conversion process highlights the importance of time and uncertainty. One of the meanings of the passage of time is the accumulation of information, in the sense that an uncertain outcome in one period is revealed in a subsequent period. This is important for decision-making under uncertainty, because a time path must be chosen in the first period with only probabilistic beliefs as to the resulting positions in the later periods, while the actual decision-making framework (or 'state of nature') applicable to later periods will only be revealed in those periods. That is, in decision theory, information accumulates over time in the sense that outcomes of random variables, affecting the decision-maker's framework, are revealed, and beliefs as to the future are better defined.

Therefore, placing the problem of global biodiversity into a sequential decision-making framework also places the role of information accumulation at the core of biodiversity. In a dynamic framework, halting the conversion process equates with purchasing time and information. This makes the expected value of this information one of the fundamental issues to be considered in determining when to halt the conversion process.

Sequential decision-making regarding resource conversions implies the passage of time, and one component of time is the accumulation of information. Since the sequence of events (i.e. passage of time) in this model is linked to decision-making regarding the conversion of the marginal state's biological resources, one clear trade-off is the postponement of the marginal conversion in order to acquire the period's information.

MVBD – Exogenous Information (XI)

$$E[MVBD]_{XI} \equiv \left\{ \frac{\partial E\left[\prod_G^{MS}\right]}{\Delta I} - \frac{\Delta E\left[\prod_G^{(MS+1)}\right]}{\Delta I} \right\} \frac{\Delta I}{\Delta t} \qquad (6.7)$$

where : $\dfrac{\Delta I}{\Delta t}$ $\equiv$ the addition to the information set overtime

Equation (6.7) incorporates one definition of time; i.e. the process by which relevant information arrives at the decision-maker. Then, information may be seen as eliminating relevant uncertainties with its arrival.

This value of biological diversity is unambiguously positive if two conditions are met: (1) information relevant to decision-making does in fact arrive by reason of an exogenous process occurring over time; and (2) the conversion of the marginal state's resources reduce the dimensionality of the gross biodiversity vector {Y}. These conditions guarantee non-negative value because an irreversible narrowing of the choice set over time (in terms of reductions in the dimensionality of the gross biodiversity vector) renders useless emerging information which would otherwise be valuable in the decision-making process. Information is valuable, but only if the choices that it implies remain available.

The Value of Exogenous Information (XI) - The Option Value of Biodiversity

$$E[MVBD]_{XI} \geq 0 \qquad (6.8)$$

$$\text{if}: (4A) \quad \frac{\Delta I}{\Delta t} > 0$$

$$(4B) \quad Y^{MS+1} \subset Y^{MS}$$

where: $Y^{MS} \equiv$ the gross set of global biological diversity with marginal (unconverted) state *MS*

This makes clear what the concept of *option value* means in the context of biological diversity. If represents the value of retaining the larger choice set until the next period's information arrives. It is, strictly speaking, the *value of flexibility* in sequential decision-making, or the *expected value of information* (Conrad, 1980). Its value in this context is clearly positive.

This result differs from much of the literature on option value, which is inconclusive as to the sign of option value (Johansson, 1987). The unambiguity of the result in (6.8) is derived from two differences in the analysis. First, this analysis focuses on a global process represented as a sequence of restrictions of the global choice set.

That is, this model presents the problem of global biodiversity as a sequential narrowing of the choice set with regard to the methods of production ('species') available for capturing biological product. If two distinct sets have an equal number of different elements (or two production vectors have common dimensionality but distinct components), then 'option value' (the value of retaining one rather than the other) is indeterminate *ex ante*. However, the problem of global biodiversity is best represented as a narrowing of the entire global choice set (of the available methods of production) rather than a substitution between elements. Then, the fundamental reason for indeterminancy is removed.

The primary reason that option value is clearly positive in this analysis is attributable to the specificity with which that term is used here. Here the 'option value' of biological diversity refers only to the dynamic values flowing from *diversity*, and this excludes many of the other values of diverse biological resources. If these other values were included, then the marginal value of non-conversion might very well be negative.

That is, option value is sometimes equated with the *marginal value of biological diversity*, as that term is defined in equation (6.3) above. However, the MVBD may be sub-divided into two further components, the appropriable and non-appropriable flows indicated in (6.4), and option value concerns only a part of those non-appropriable flows found in (6.4B). Specially, 'option value' concerns only the values within (6.4B) that flow from the existence of a dynamic aspect to the problem.

Therefore, the change in global values from marginal conversions may be segregated into three distinct categories: (1) the difference in the value of appropriable flows of biological resources; (2) the static (portfolio) non-appropriable value of diversity; and (3) the dynamic (expected value of information) non-appropriable value of diversity. Many times, the term 'option value' has been applied to the aggregation of some combination of these three effects (yielding ambiguous results), but here it is used only in regard to the values derived by reason of the use of a dynamic decision-making framework. That is, option value concerns only the third category of diverse resource values listed above.

Option value is thus restricted here to mean only the value to be derived from retaining flexibility in a sequential decision-making framework. With the arrival of information over time, the value of this retained flexibility must be positive. In this sense, the option value of biological diversity must always be positive.

It is logical to use this more restrictive definition in regard to the option value of biodiversity. The first two categories of value listed above are static values, i.e. they would flow from decision-making even in the context of a one-period framework. On the other hand, the expected value of information (or 'option value') applies only if there is a future. The value of such information will ultimately flow to the global community in terms of either increased mean yields or reduced variability; however, retaining these options has value even in earlier periods, to the extent that it is expected that their retention will be important. Diversity is retained in order to provide a dynamic as well as a static form of insurance. It is only important to retain options to the extent that it is anticipated that 'information' will arrive from the environment to make these options important for future decision-making.

It remains to explain why it is that information will necessarily arrive in this particular decision-making process. In sequential decision-making, information is the occurrence of non-deterministic change in the decision-making environment. That is, information emerges between periods if there is some relevant alteration in the environment that cannot be predicted with certainty; the passage between decision periods reveals the 'state of nature' which could otherwise only be probabilistically projected.

The nature of the biological world assures precisely this result. It is the very essence of a dynamic system, in which the processes of mutation, selection and dispersal continuously alter the natural 'state of nature'. In regard to small organisms, such as bacteria, viruses, and insects, these biological processes can occur very rapidly, literally reproducing thousands of generations in a single year. The biological process is random and evolutionary, not deterministic, and to the extent that it can be understood, it is too complex to predict.

It is the continuous state of motion within the biological world which guarantees that time produces relevant information. The nature of this information in a biological world is the extent of the shifting of the human niche. The insurance that we have for adapting to such shifts is the diversity of the species upon which we rely, or upon which we might rely. Marginal conversions represent losses of such options. The expected marginal value of biological diversity includes as a component the expected value of receiving information prior to the foreclosure of options. The value of this component is clearly positive.

6.4.4 Dynamics: The Value of Endogenous Information

It was noted above that the intra-state variability of a larger vector of resources was likely to be smaller, and that states in the group South were the more likely states to rely upon a larger vector of resources. In fact, it is also likely that these states will also use biological assets with greater inherent variability in yield (and value). This is because, assuming regular concave individual utility functions, specialised production probably will be directed toward high-mean/low-variability biological assets. The variability occurs in the aggregate, because so many individuals specialise in precisely the same assets. In fact, from many years of use, the individual distributions of the yield from specialised biological assets will be very well-known, implying a very small amount of information to be gained from their use.

Unused, and less-used, biological assets are relatively unknown quantities. The expected variability of any individual non-specialised biological asset is quite large, especially relative to its expected mean, precisely because so little is known about it. The expected information acquired from the exploration of these commodities is much higher than it is with the specialised assets, and this *exploration value* is positive.

There is therefore, a differential informational value to the exploration of the non-specialised biological assets that derives directly from the fact of their relative obscurity. This is termed the *expected value of endogenously generated information.*

MVBD – Endogenous Information (NI)

$$E[MVBD]_{NI} \equiv E\left\{ \frac{\Delta\prod_G^{MS}}{\Delta I} - \frac{\Delta\prod_G^{(MS+1)}}{\Delta I} \right\} \frac{\Delta I}{\Delta y} \frac{\Delta y}{\Delta R} \frac{\Delta R}{\Delta t} \tag{6.9}$$

Equation (6.9) states that non-conversion not only maintains the choice set at its differentially greater size, it also maintains the information flow at its differentially greater rate. The marginal conversion reduces the potential information set and the potential choice set in a single act. The reduction of either alone is costly in a sequential decision-making framework. Therefore, the expected value of endogenous information is also invariably positive, being a mere sub-set (but a very important sub-set) of the entire information flow occurring between decision points.

6.4.5   Conclusion – The Value of Global Biological Diversity

The value of global biological diversity has been given a very specific definition in this section. It is the global impact of the conversion of the marginal state to specialised biological resources. This global impact has four components, two static and two dynamic (i.e. information-based). The information-based values of biodiversity will ultimately feed through the static components; however, in any given period there is also the expected value of this future flexibility.

The Components of the Value of Global Biological Diversity

$$E[MVBD] \equiv (A)\frac{\partial\prod_G}{\partial(y^d - y^{sp})} + (B)\ PE + (C)E[XI] + (D)\ E[NI] \tag{6.10}$$

As discussed in this section, the sum of the values of terms (B), (C) and (D) in equation (6.10) is strictly positive in the context of the marginal state's decision whether or not to convert its diverse resources. However, it is also clear that the value of term (A) has been pronouncedly negative over many years, inducing successive states to undertake resource conversions.

It is not really possible to say anything about the relative magnitudes of these two forces. The force for conversion, term (A), flows through a vector of outputs which are clearly appropriable by the host state. The values of biological diversity, on the other hand, are clearly non-appropriable values. The value of global variability reductions is a pure public good; it is not possible to exclude any individual or any state from receiving its benefits. The other values are informational in nature, and these are very diffusive and inappropriable.

Therefore, in comparing these two forces, one for and one against conversion, their most distinguishing feature is their degree of appropriability. Under existing institutions, it is clearly not possible to gauge the relative magnitudes of these two forces at this point in the global conversion process. However, this does not imply that there is no value to biodiversity, only that it is not being given effect.

6.5 THE INTERNATIONAL REGULATION OF EXTINCTION: THE GENERAL PRINCIPLES

The nature of the global biodiversity problem is one of imperfections in the decentralised (i.e. multinational) regulation of the conversion process, because individual states do not take into consideration the consequences of their conversion activities on global stocks of diverse resources. The obvious solution to the global biodiversity problem is to halt the global conversion process at its global optimum, taking into consideration these stock-related externalities.

This 'solution' is also a regulation problem, as it must also be implemented through the decentralised process that created the biodiversity problem. The agents with control over the conversion process are the individual host states, under the doctrine of national sovereignty, and the global community must take action regarding conversions through these agents. Therefore, the nature of international regulation regarding conversion is necessarily the construction of *international incentive mechanisms*, devised to induce individual states to regulate conversion in accord with global objectives.

The fundamental purpose of these incentive mechanisms must be to bring the global effects of diverse resource conversions within the decision-making framework of the marginal states. That is, as discussed in section 6.3, the object is to internalise the externality, causing individual states to recognise the true opportunity costs of diverse resource conversions. In essence, the incentive mechanism should have the effect of bringing the MVBD (the marginal cost curve in Figure 6.1) within the decision-making framework of the marginal states.

The remainder of this section considers the general nature of this intervention, where the assumption is that the object of the global community is indirectly to implement the optimal stopping rule that cannot be implemented directly. That is, the objective is to halt conversions at the global optimum, given the globalised values of biodiversity, in a world in which there is little information on the latter and no inherent capacity to do the former.

6.5.1   International Intervention in National Resource Management

Individual states which decide to convert their resources do so through a set of investment decisions. In essence, states choose to invest limited national funds in the management and production of some species, and not in others. It is the presence or absence of investment that determines the fate of a particular species.

The decision regarding conversion to the use of modern agriculture has become embedded in a small slate of specialised species. Therefore, most available investment funds will be absorbed by these species alone, since the decision to invest in them demonstrates the decision to convert to capital-based biological production.

The role of international intervention is to induce the inclusion of diverse habitats and diverse/non-specialised species within the investment portfolios of the marginal unconverted states, thereby halting the global conversion process. This is accomplished by means of optimal investments by the global community in management institutions that will induce these investments by host states. That is, the role of the international community is to alter the incentives for states themselves to create better incentive systems for the individual management of diverse biological resources. It is a new institutional framework overlaying what was previously an exclusively domestic regulation problem.

In the *global model of bio-institutional regulation*, in equation (6.11) below, the regulatory structure existing in the absence of international intervention is captured in terms (6.11b), (6.11c) and (6.11d). In (6.11a), the state regulator simply acts to maximise the joint benefits to the collectivity of harvesters of the resource, and does so by investing in the institutional structure regarding the resource ($M(M_s)$ in (6.11b) and

(6.11d). The bio-institutional structure (6.11d) then determines how individuals will fare in their interdependent production of this resource, and thereby determines in part how these individuals will themselves invest in this resource. Therefore, overall investment in the diverse resource, M, is a function of the initial state spending on this management, M_s.

This model then demonstrates how action by the global community has impact at the ground level in the Southern states. The intervention takes the form of optimal investments in management (M_G) undertaken with the object of inducing state-level investments. When successful, these investments then determine the rate at which the host state will invest in that resource (equation 6.11(b)), and the aggregate rate of investment determines the bio-institutional structure within which the resource is produced (6.11(d)). This structure then determines the management of the resource of individuals operating in that state (6.11(c)).

The Global Model of Bio-Institutional Regulation:

Global Regulator's Objective

$$M_G^\Omega : \frac{\text{Max}}{M_G} \prod_G \equiv \int_{t=0}^{\infty} e^{-rt}\left(\prod_s (y^{d\cdot};M_G) + \prod_D (y^{sp};x,R) - r\rho_M M_G\right) dt \quad (6.11a)$$

where

$\prod_G$	$\equiv$	the global benefits from biological production
$\prod_S$	$\equiv$	the benefits to nonconverted states
$\prod_D$	$\equiv$	the benefits to converted states
M_G	$\equiv$	the management expenditures by the global community

Subject to:

State Regulator's Objective

$$M_S^*(M_G) : \frac{\text{Max}}{M(M_S(M_G))} \int_{t=0}^{\infty} e^{-rt}$$

$$\left[\sum_{i=\mu}^{n} \prod_i (y_i(M), R_i(M), M_i(M); x(M), R(M), M) - r\rho_M M_S\right] dt \ \forall \ s \quad (6.11b)$$

Subject to:

Interdependent Individual Choices

$$y_i^*(M), R_i^*(M), M_i^*(M): \quad \underset{y_i, R_i, M_i}{\text{Max}}$$

$$\int_0^\infty e^{-rt}\left[p(y^d)y_i^d - c(R,M)y_i^d - r\rho_R R_i - r\rho_M M_i\right] dt \ \forall \ i \qquad (6.11c)$$

Subject to:

Bio-Institutional Structure

$$i = \mu(M),\ldots\ldots, n \quad \text{where} \quad 1 \leq \mu(M(M_S(M_G))) \leq n; \mu' > 0, \mu'' < 0$$

$$\dot{x} = H(x,R) - y \qquad (6.11d)$$

In the absence of international intervention, the outcome of this decision-making process often will be non-management on the part of the state, and overexploitation on the part of the harvesters. With optimal international investments there is the possibility that additional diverse resources will be included within the investment portfolio of the host states, and thus the global community.

The general nature of the global community's institution-building is evident. It must invest in management institutions that will cause the values of diverse resource stocks to be channelled through the hands of the host states. In section 6.2, the problem of global biodiversity losses was described in relation to the definition of the vector y^d: the appropriable flows from diverse stocks of biological resources under existing institutions. At the core, the economic problem concerns the costliness of institution-building for the internalisation of the other benefits from these stocks. As re-channelling these otherwise inappropriable benefits will be a costly activity, it will require a commitment of resources to bring this about.

The optimal level of international intervention for this purpose will be a trade-off between the effectiveness of these investments (in terms of the enhanced benefits from increased retention of diverse stocks) against their opportunity costs (in terms of forgone investments). With regard to any given diverse resource stock x and base R, the international community will invest in its maintenance to the extent that its marginal benefit equates with its marginal costliness (equation 6.11(a)). The net benefits from increased management expenditure will be the sum of the increased appropriable benefits flowing through the host state (by reason of enhanced appropriability or increased productivity) and the increased non-appropriable benefits flowing to the global community (by reason of enhanced stocks).

Optimal Global Investments in Biodiversity

$$M_G^\Omega : \frac{\partial \prod_S}{\partial y^d} \frac{\partial y^d}{\partial M_G} + \left[\frac{\partial \prod_D}{\partial x} \frac{\partial x}{\partial M_G} + \frac{\partial \prod_D}{\partial R} \frac{\partial R}{\partial M_G} \right] = r\rho_M \quad (6.12)$$

Therefore, it is the global community's role to attempt to bring a stopping rule into effect in the global conversion process. It will do this by investments in incentive mechanisms that are intended to induce a change of investment portfolios by states that would otherwise convert their biological resources. The general nature of these systems must be an enhanced channelling of the stock-related benefits through the hands of the host state.

6.6 CONCLUSION

The global problem of biological diversity losses may be viewed as the problem of halting the global conversion process at the globally optimal point in time (and space). Individual state decision-making cannot achieve this because there is no recognition of the countervailing benefits to the force for conversion. The marginal value of biological diversity contains a component of non-appropriable benefits that flow to the global community at large.

There is no doubt that there is a value to diversity. The decline of global stocks of diverse resources necessarily results in losses of global insurance and information services. There are services that flow to the entire global community, and this is the fundamental reason for their importance and for their decline.

It is only international intervention that can halt this conversion process. However, the international community can act on this problem only through the instrument of the host state. The nature of international action must be inducement of an otherwise non-existent pattern of investments by the Southern states.

The global solution to the global biodiversity problem is investment in international institutions by the global community. These investments must establish mechanisms which provide incentives for host states to invest in diverse resources. The nature of these incentives will be the re-channelling of the global values of diverse resources back to the host states themselves. Such a system of intervention provides the necessary incentives to halt the global conversion process at the optimal point, while maintaining the capability of the marginal states to pursue development (albeit down diverse pathways).

Therefore, the global diversity problem may be viewed as the externality emanating from the pursuit of the same, narrow pathway to development by successive states. As the final states embark on this same avenue, the commonality in the methods of production employed jeopardises the biological production process itself. The solution must then be the encouragement of the remaining uncommitted states toward diverse pathways of development. However, global institutions for development have always been constructed around the ubiquitous model, and they have ignored the alternatives. The global biodiversity problem requires the construction of international institutions focused upon alternative development routes.

7 International Intervention in National Resource Management: The Instruments for Regulating Global Biological Diversity

7.1 INTRODUCTION

The global problem of global biological diversity losses is the result of decentralised (state-based) decisions concerning land use and conversions. Each state acting independently will take no account of the value deriving from the maintenance of a globally diverse portfolio of biological resources. Instead, each will attempt to specialise its own portfolio to its own individual benefit. This decentralised decision-making process will result in the targeting of a globally suboptimal stock of diverse resources.

The object of global regulation is to regulate the conversion process so that these global stock effects are taken into consideration. In theory, this requires only that each decision on conversion be made to balance the marginal benefits against the marginal costliness (including the stock effects) of that decision.

This chapter concerns the nature of the instruments required to implement this policy for the global regulation of extinctions. Although the policy is easily stated, its implementation is far more complex. This is because terrestrial resources are national resources; they are managed exclusively by an independent, sovereign state. International intervention must operate through the decision-making framework adopted by that state.

Therefore, this chapter constitutes an enquiry into the fundamental nature of the *international environmental agreement* required to address the problem of global biodiversity. It takes the multinational nature of diverse resource regulation as a given, and then analyses the nature of the international agreement that will enable cooperation between these various states to be effected on this subject.

It is an important and timely subject precisely because so much effort is currently being undertaken on just this task. The United Nations Conference on Environment and Development adopted the text of a Convention on the Conservation of Biological Diversity in June of 1992, and the next few years will give concrete meaning to this convention through various resolutions and protocols. In addition, the Group of Seven states initiated another institution for the purpose, in part, of conserving global biological diversity in 1990: the Global Environmental Facility. This is another example of an international institution for biodiversity conservation that is currently taking shape.

The remainder of the discussion within this book provides an analysis of the requisite characteristics for an effective international agreement on biodiversity conservation (assuming that the problem is of the nature described in the preceding chapters). In this chapter the general nature of such an international agreement is discussed. Then, in subsequent chapters, the specific components of that agreement are detailed.

The obvious object of an effective international agreement is to alter the choices of the 'host states' with regard to their diverse resources. So, for example, with regard to the four African states that failed to invest in the management of their elephant populations in the 1980s, resulting in the loss of half of the continent's total population, the objective is to induce changes in their spending decisions. In general, *effective* global management of diversity will usually be fully translated into effective domestic management of diverse resources; the problem is to devise a system that will convert global investments into domestic investments in diverse resources.

This is, however, exactly half of the environmental problem of biological diversity. The other half of the problem lies in the inducement of optimal transfers from the 'Northern' states in compensation for the flows from stocks of diverse resources. There can be no solution to the other half of the problem without addressing this half as well. The core of the problem lies in the suboptimal flow of payments for these intangible services, and increasing that flow away from zero (although moving in the right direction) does not assure that the resolution is in the neighbourhood of an efficient solution. For that to occur, there must be equal attention paid to the question of inducement mechanisms for the North.

International intervention in national management decisions is always a difficult problem, but when the values involved are as complex as the informational and insurance services of biological diversity, the degree of difficulty rises to a higher order. This chapter considers the fundamental nature of the solution to this problem of international intervention for the management of biodiversity.

7.2   THE NATURE OF THE PROBLEM – OPTIMAL BIODIVERSITY POLICY

The regulatory problem of global biodiversity losses concerns the optimal means of inducing specific forms of investments by sovereign states. The extinction problem results from discrete decisions made by independent states concerning their individual investment portfolios. Each state has a limited amount of resources to invest, and it must therefore restrict itself to a finite number of productive assets to hold within its portfolio. Biological assets, i.e. species, are one form of productive asset that the state may choose to hold. Extinctions result when states choose to invest in assets other than specific biological ones, because the returns on the biological assets are inferior to others.

The specific decision problem faced by the state was developed in equation (5.15). The state's decision concerning a given resource devolves to the determination of how much state funding should be dedicated to its management (M_s). The outcome of this decision will determine the quality of resource exploitation in that state; large spending on management results in efficient rates of exploitation, lesser amounts introduce increasing degrees of overexploitation and waste. In effect, the absence of adequate management spending by the state creates an environment that is not conducive to individual investment in the resource.

Individuals living in and near the natural habitat of the resource respond with low levels of investment in the resource (its stock level), its habitat (e.g. the forest), and its monitoring. Their decisions are responsive to the state's in the sense that they are based on beliefs about the future that are determined by the state's management spending. There is little reason to invest in resources that the state leaves unsecured.

The state's decision-taking in regard to specific resources is in turn the consequence of expected investment-worthiness. Some biological assets are viewed as unworthy of substantial investments on account of relative growth rates (H), relative unit rental value (λ), and even relative management costliness (M). The global extinction problem arises from the fact that, at present and in recent times, a small number of biological assets have attracted the vast majority of societal investments in biological production, leaving the great majority in an underinvested and overexploited state.

The global community's objective is to create the optimal investment portfolio, considered from the global perspective. In order to accomplish this, it must alter the existing investment portfolios of individual states. This is accomplished through investments (by the global community) which induce consequential investments by host states in particular biological assets. Therefore, in solving the international problem of terrestrial extinctions, the global community must act through individual sovereign states, inducing them to alter their investment portfolios in the global interest.

The specific nature of the global community's decision problem is set forth in equation (6.11). The global community decides upon an investment portfolio in biological assets which maximises the value of the portfolio to the global community. The global community then acts by means of investments in global management mechanisms (M_G) which operate through their impact on state management decisions (M_s). State management decisions consequently determine state investment portfolios (and the global portfolio in the aggregate).

In short, the problem of extinction is an example of the problem of inefficient decentralised management of a global resource. Individual states are taking decisions regarding their individual portfolios without adequate consideration of the effects of these decisions on global stocks of diverse resources. The global community's role is to act as an 'international regulator', taking actions to induce changes in individual state's portfolios. The capacity of the global community to intervene in state decision-making is the subject of this chapter.

7.3   OPTIMAL BIODIVERSITY POLICY WITH INSTRUMENTS

The above discussion makes it clear that the primary objective of international intervention is to induce state expenditures on the management of particular diverse resources. Changed management expenditure will then be reflected in altered investment portfolios. This means that the overarching object of the international regulation of extinction may be expressed as the maximisation of the difference between the benefits (flowing from an altered global portfolio) and the costs (of global management to induce changed domestic investment patterns). This objective is depicted in equation (6.11a).

The cost-effectiveness of global investments in biodiversity management will be dependent upon the quality of the instruments used to implement them. That is, the global objective will need to be maximised subject to the available instruments for influencing state investments.

Available Instruments

$$M_s = I_G(M_G^I) \tag{7.1}$$
$$\dot{M}_G = T_G(M_G^T)$$
$$M_G = M_G^I + M_G^T$$

where : M_G^I ≡ global management invested in inducing host-state investment

M_G^T ≡ global management invested in inducing North–South transfers

I_G ≡ the incentive system for inducing M_s

T_G ≡ the incentive system for inducing M_G

The global objective is then to maximise (6.11a) by the optimal choice of M_G^T, M_G^I as defined in (7.1). That is, the global objective requires the global community to choose the optimal instruments and investment levels for conserving global biological diversity. Then (6.11a) may be rewritten as follows.

Global Objective (Optimal Investments and Instruments)

$$M_G^\Omega : \frac{\text{Max}}{M_G, M_G^I} \int_0^\infty e^{-rt} \left[\prod_G (M_s(M_G^I)) - r\rho_M M_G \right] dt \tag{7.2}$$

where : $\prod_G$ ≡ net value of altered investment portfolios to global community

M_S ≡ management by states of group South of diverse resources

M_G ≡ management by global community of diverse resources

ρ_M ≡ price of management services

r ≡ the opportunity cost of capital

Equation (7.2) shows that the global community intervenes through investments in mechanisms which will induce state investments in diverse resources. The trade-off here is between the effectiveness of those (global) investments and their costliness. This rendition of the global objective makes clear that a sub-object of the global community is to construct *two* optimal instruments for the maximisation of (7.2). One of these instruments (I_G^*) will be used to obtain the most cost-effective inducement of optimal supply state investments in the diverse resources. The other instrument (T_G^*) will be used to obtain the most cost-effective inducement of optimal demand state transfers for diverse resource services. Their cost-effectiveness will be determined in regard to their capacity for generating global benefits with the least amount of aggregate investment (M_G).

Therefore, the regulation of extinction involves the creation of incentive systems for both halves of the world. This is because the nature of the regulatory problem is reciprocal. The states in group North (the already-converted states) have the incentive to free-ride on the provision of diverse stocks (x) and habitats (R), and the flows that they imply. The existence of these uncompensated flows then leads to the problem of state underinvestment in diverse resources, and hence global extinctions. However, should the states in group North decide to transfer funds to group South in order to compensate for these flows, then the incentives to free-ride are reversed, not eliminated. That is, then the states in group South have the incentive to keep the funds, but apply the payments to uses other than diverse resource management (investing in the first-best portfolio from their own perspective).

A reciprocal incentive system is required in order to solve this problem. One part of the system must focus upon the inducement of *optimal* transfers from North to South (i.e. from converted to unconverted states). The other part must focus on the inducement of the selection of the *globally* optimal investment portfolio by the Southern states. The remainder of this chapter discusses the means for accomplishing the latter objective; the next chapter discusses the possibilities for pursuing the former.

7.4   THE NATURE OF INTERNATIONAL INTERVENTION

How is it possible for 'global' spending to be translated into domestic management? For example, with regard to the four African states whose unofficial 'open access' regimes resulted in the halving of the continent-wide elephant population, how could spending in one part of the world put additional patrols in place in another part of the world? This is a crucially important piece of the regulatory problem for the effective regulation of global terrestrial resources.

This section analyses the impact that different forms and patterns of 'global' spending will have on domestic investments in diverse resources. For this purpose it is necessary to construct the 'reaction function' of the owner-state, in order to determine how intervention will affect state choice. Any intervention must operate through the supplier state's decision-making framework regarding the worthiness of investments in diverse resource management; this framework is described in equation (5.15). It implies that owner-states will invest in the management of diverse resources to the extent that the resources render benefits sufficient to pay for that management. This implies the following optimality condition for state management decisions; owner-state investment in management is simply a function of the marginal benefit from enhanced co-operation in the exploitation of that commons.

State Investment in Diverse Resource Management (w/o intervention):

$$M_s^* : \frac{\partial \Pi_s}{\partial M_s} = r\rho_M \tag{7.3}$$

where : Π_s ≡ the aggregate value to citizens of states from a diverse resource

≡ $\sum_i (\lambda_s y_i^d)$ where λ_s is the unit resource rental value in state s

State investments in the management of a resource yield some level of return, by reason of the increased levels of co-operation in the exploitation of the resource and the consequent movement toward first-best management. Within this framework, management is a factor of production with regard to the diverse resource, with non-decreasing but diminishing returns.

In the absence of global intervention, no state will invest to achieve first-best management, on account of the costliness of such investments. Therefore, the marginal benefit of further management spending on diverse resources will remain positive at the state's 'local' optimum (due to the marginal costliness of that spending); this result is demonstrated in question (7.4).

The Marginal Benefit from State Investments in Diverse Resource Management (Without Intervention)

$$\frac{\partial \Pi_s}{\partial M_s} \equiv \frac{\partial \Pi_s}{\partial \lambda_s} \cdot \frac{\partial \lambda_s}{\partial \mu} \cdot \frac{\partial \mu}{\partial M_s} > 0 \tag{7.4}$$

because $\dfrac{\partial \Pi_s}{\partial \lambda_s}$ > 0 from equation (7.3)

$\dfrac{\partial \mu}{\partial M_s}$ > 0 by definition (see equation 5.15)

$\dfrac{\partial \lambda_s}{\partial \mu}$ > 0

$$\text{Since:} \quad \lambda_s = p \cdot \left[1 + \frac{1}{\varepsilon_d (n - \mu)} \right] - c$$

$$\longrightarrow \quad \frac{\partial \lambda_s}{\partial \mu} = (-1)(-1) p \frac{1}{[\varepsilon_d (n - \mu)]^2} > 0$$

Equation (7.4) also depicts the process by which state investments in management yield benefits within that state. Management is effective by means of the removal of uncooperative access to the resource, so that when additional resources are devoted to management the level of uncooperative exploitation decreases (i.e. $n-\mu$, the non-co-operative fringe, decreases). Better cooperation in the exploitation of the resource in state s results in better per unit rental value for each individual within state s harvesting the resource (λ_s increases). And this consequently results in higher aggregate benefits from the resource within the state. So long as there is increased cooperation from increased management expenditures (even though the returns may be diminishing), there is a positive return to state investments in management.

Nevertheless, these returns may not be sufficient to warrant investment by the state in certain diverse resources. The state's decision problem (set forth in equation 5.15) requires that investments in diverse resource management be competitive with the returns available from other assets within the economy. The failure to meet this condition dooms many biological resources.

In order to alter this result, the global community may intervene by way of institution-building that results in the enhancement of the returns to state investments in diverse resource management; this is the decision problem set forth in equation 6.11. In essence, the global community, through its investments, develops another flow to the supplying state's investments, in addition to that shown in equation (7.4). The one unit of state management spending will generate two distinct flows of benefits: one sourced in the owner-state's internal production system and one sourced in the global community's external support institution.

The Impact of State Investments in Diverse Resources (w/intervention)

$$\frac{\partial \prod_s}{\partial M_s} (M_G) : \left(\frac{\partial \prod_s}{\partial M_s} \right)_s + \left(\frac{\partial \prod_s}{\partial M_s} \right)_G \tag{7.5}$$

where : $(\cdot)_S$ ≡ the (state sourced) benefits derived from state
                    management (as in eq. 7.5)
         $(\cdot)_G$ ≡ the (global community sourced) benefits derived
                    from state management

Equation (7.5) gives the basis for a reaction function for the owner-state. Optimal state investment decisions (as determined in Equation (7.3)) will now generate new levels of state investment spending on diverse resources *dependent on the level of investment by the global community, i.e.* $M_s^*(M_G)$.

The effect of intervention is to enhance the perceived effectiveness of state investments in the particular diverse resource, thereby enhancing the incentives to make those investments. The global community does this through the construction of a mechanism that automatically responds to state investments with benefit enhancements. Those investments then determine the likelihood of the maintenance of the diverse resources within that state.

7.5 ALTERNATIVE INSTRUMENTS FOR INTERNATIONAL INTERVENTION

International intervention (in order to induce the selection of the global optimum over the local) may take one of three general forms:

(1)   (I_1): direct inducement – the inducement of changes in the investment portfolio of the supplier state by means of direct payments for commitments to conserve diverse resource stocks (x, R).

(2)   (I_2): indirect inducement (rent appropriation) – the inducement of changes in the investment portfolio of the supplier state by means of market regulation in the consumer state for the purpose of maximising the appropriation of rents from diverse resource flows by the supplier states.

(3)   (I_3): indirect inducement (flow appropriation) – the inducement of changes in the investment portfolio of the supplier state by means of investments in market regulation in the consumer state for the purpose of channelling the flow of benefits from diverse resources initially through the supplier states.

The distinction between the first and the latter two methods lies in the variable to which the global community responds. Direct inducement operates by enhancing supply state benefits in accord with retained stocks of diverse resources. Indirect inducement operates by means of enhancing supply state benefits that flow from diverse resource stocks.

The distinction between these three routes for intervention is formalised in equation (7.6). Here, the benefits to state *s* from a diverse resource flow (y^d) is broken down into three distinct components. First, there are the stock-related benefits (deriving from the mere existence of stocks *R*, *x*). Secondly, the flow of benefits is segregated into the appropriable component (y^{da}) and the non-appropriable (y^{dn}); the former earns a return equal to the local rental value (λ_s). Third, the non-appropriable flows generate no locally appropriable returns. Intervention may occur by means of the attempted enhancement of the returns from any one of these three components.

Alternative Routes for Intervention:

$$
\begin{aligned}
\prod_s (y^d) &= M_G^{I_1}(R,x) + \lambda_s(M_G^{I_2}) \cdot y^{da}(M_G^{I_3}) + 0 \cdot y^{dn}(M_G^{I_3}) \\
&= M_G^{I_1}(R,x) + \lambda_S(M_G^{I_2}) \cdot [y^d - y^{dn}](M_G^{I_3}) \qquad (7.6)
\end{aligned}
$$

where : $y^d = y^{da} + y^{dn}$

$y^d \equiv$ the flow of services from a diverse resource

$y^{da} \equiv$ the appropriable flow (under existing institutions)

$y^{dn} \equiv$ the non-appropriable flow

$M_G^{I_1} \equiv$ the extent of international intervention via instrument I_i

As indicated in equation (7.3), the level of management spending by state *s* on the diverse resource is a function of the aggregate state benefits expected to be received from that resource. International intervention operates by enhancing those expected benefits, and each of the available instruments does this in a different fashion. Direct inducement (I_1) operates by making transfers directly related to stocks of diverse resources (and is thus indirectly related to service flows); this instrument is the subject of this chapter. Rent appropriation (I_2) acts by increasing the per unit resource rental value received by the supply state for each unit of appropriable flow generated; this instrument is the subject of Chapter 8. Finally, flow appropriation (I_3) is directed toward shifting more of the diverse resource flow (y^d) from the non-appropriable to the appropriable vector; this instrument is the subject of Chapter 9.

The global community's investments add the second term to equation (7.3), effectively enhancing the benefits from state management spending.

The precise nature of this term depends upon the instrument used for directing benefits to the supplier state. Each instrument has the global community respond to state investments with its own investment, but the nature of the global investment depends on the instrument.

Alternative Global Investments/Instruments

(1) Direct Subsidy

$$I_1 : \left(\frac{\partial \prod_s}{\partial M_s}\right)_G = \frac{\partial \prod_s}{\partial M_G} \cdot \frac{\partial M_G}{\partial M_s} \tag{7.7}$$

(2) Rent Appropriation Mechanism

$$I_2 : \left(\frac{\partial \prod_s}{\partial M_s}\right)_G \equiv \frac{\partial \prod_s}{\partial \lambda} \cdot \frac{\partial \lambda}{\partial M_G} \cdot \frac{\partial M_G}{\partial M_s} \tag{7.8}$$

(3) Flow Appropriation Mechanism

$$I_3 : \left(\frac{\partial \prod_s}{\partial M_s}\right)_G \equiv \frac{\partial \prod_s}{\partial (y^d - y^{dn})} \cdot \frac{\partial (y^d - y^{dn})}{\partial M_G} \cdot \frac{\partial M_G}{\partial M_s} \tag{7.9}$$

There are three points to be garnered from this depiction of the alternative instruments for managing extinction.

First, it is apparent that international intervention must take the form of a benefit supplement, i.e. it responds to domestic investments with international investments which supplement the domestic investment effects with globally supplied benefits. This is depicted in equation (7.5).

Secondly, it is also apparent that international intervention must operate dynamically. That is, in each period state management spending must be induced by the expectation of a global response (dM_G/dM_s in equations 7.7 through 7.9). This is probably the reverse of the causation that is usually assumed, i.e. that global expenditures will cause domestic conservation. The latter conception is only true to the extent that a history of past responses will build up an expectation of future responses, thus inducing state investments (this will be explored further in section 7.6).

Finally, this also demonstrates once again the nature of the three alternative routes for global intervention. All three are mechanisms instituted to generate benefits supplementing state-generated benefits from diverse resource investment. The global investment induced may be of the nature of a direct subside (eq. 7.7), or of an investment sustaining an already instituted mechanism for global transfers (eq. 7.8 or 7.9). Either through the subsidy or the mechanism, the benefits of domestic investments in diverse resources are enhanced.

In sum, global intervention takes the form of global investments *contingent upon state investments in particular diverse resources*. This section has defined both the three alternative routes generally available, and the general nature of global intervention in national resource management. The following sections demonstrate how these definitions result from an analysis of the problem of inducing state commitments to alternative investment portfolios.

7.6 INTERNATIONAL INTERVENTION AS INSTITUTION-BUILDING

It is necessarily the case that global management must be perceived as responsive to state management spending in order to induce state management spending. This is so because the object of international intervention is to induce investments that are not perceived by the state to be in its own interest, i.e. in an investment portfolio that is optimal from the global but not the national perspective. Therefore, for a state to honour a contract against its own interest, there must be an enforcement mechanism for that contract. Making the global community's investment responsive creates the enforcement mechanism as part of the contract.

It is also necessary for global intervention to take the form of institution-building precisely because the reshaping of the supplier state's investment portfolio hinges upon the remoulding of that state's beliefs concerning the future. It is because states come to believe that the response to domestic investments will be enhanced benefits that their portfolios will be altered. That is, global investments in conservation do not in themselves generate any changes in domestic state investment patterns; it is instead the effect of those investments on state expectations that matters most. It is the role of institutions to build and fulfill expectations.

This may be demonstrated with a simple hypothetical example. Assume that the global community initiates its flow of enhanced benefits (M_G) in return for the domestic regime's commitment to correspondingly increase conservation spending (i.e. $dM_s/dM_G = 1$). That is, the global community promises to invest M_G in that state each period so long as the state commits these funds to conservation. Therefore, the state is promised a future subsidy so long as it adopts the constraint $M_s = M_G$ in current decision-making, but once the state adopts unconstrained decision-making the subsidy goes to zero. In equation (7.10), the adoption of constrained decision-making is indicated by the presence of a semi-colon in the argument of the state's benefit function. The expected value of the future subsidy is equal to the present value of the flow of the promised subsidy (discounted by the 'credibility' of the global community in honouring its promises, v).

Even if the owner-state gives its commitment to conserve biodiversity, the optimising owner-state will honour this commitment (and thus alter its investment portfolio) only if the expected *future benefits* outweigh the immediate benefits from investing in the locally perceived first-best investment portfolio.

Time Consistency of Owner-State's Commitment to Diverse Resource Management

Honour Commitment to Invest $M_s=M_G$ in Diverse Resources iff

$$\prod_s (M_G; M_s = M_G) + v \frac{\prod_s (M_G; M_s = M_G)}{(1+r)} + v^2 \frac{\prod_s (M_G; M_s = M_G)}{(1+r)^2} + \ldots$$

$$> \prod_s (M_G) + \frac{\prod_s (M_G = 0)}{(1+r)} + \frac{\prod_s (M_G = 0)}{(1+r)^2} + \ldots \ \forall \ t$$

$$\longrightarrow \frac{v}{(1+r)-v} \prod_s (M_G; M_s = M_G) - \frac{1}{r} \prod_s (M_G = 0)$$

$$> \prod_s (M_G) - \prod_s (M_G; M_s = M_G) \ \forall \ t \qquad (7.10)$$

where: $\prod_s(\cdot)$ = the benefits from enhanced unconstrained investment

$\prod_s(\cdot;\cdot)$ = the benefits from enhanced constrained investment

v = the expected likelihood of receipt of global investment

Equation (7.10) formalises the conditions under which the owner-state will honour its commitment. The LHS of the last equation represent the present value of the expected benefits from honouring the commitment (and assuming the portfolio constraint) while the RHS of this equation represents the present value of the costliness of the assumption of this portfolio constraint. The net benefits correspond to the expected value of the stream of payments from the global community; the net costliness corresponds to the difference for the owner-state between constrained and unconstrained optimisation. The state's commitment to this constraint is time-consistent only if its expectations justify its adoption in each and every period (Kydland and Prescott, 1977).

In this sense, all interventions by the global community must be dynamically constructed. Since there is no explicit external enforcement mechanism for a contract between sovereign states, the contract between the states must itself provide this mechanism. This is not an insuperable obstacle, as many if not most private contracts also provide for internal enforcement mechanisms. However, the creation of internal enforcement mechanisms for international commitments represents institution-building, and the solution to the international extinction problem requires institution-building of this sort.

The nature of the institution-building required to solve this problem is indicated by a closer inspection of (7.10). Again, the first term on the LHS represents the expected flow of benefits from enhanced global investments resulting from current domestic portfolio changes. This term is discounted by the perceived likelihood that the global community will make future contributions (v). Here it is assumed that the state's expectations have a 'knife-edge' character to them, i.e. the global community loses all credibility (for all time) when it fails to honour its obligation just once but retains some probabilistic level of credibility (here assumed constant for simplicity) when it performs its obligation.

The motivation for institution-building is captured in this term. Consider initially the nature of the global community's intervention if credibility is not a matter of concern; that is, in the first instance, assume that there is perceived certainty of future benefits. Then states will invest in diverse resources (against locally prevailing incentives) to the extent that these investments generate enhanced benefits (from expected future global investments) at a rate that meets the existing rate of return on productive assets.

Asset Selection – Consistency of Commitment to Conversion (with v=1):
Honour Conservation Commitment iff

$$\prod_s (M_G; M_s = M_G) - \prod_s (M_G = 0) > r \cdot \left(\prod_s (M_G) \right.$$

$$\left. - \prod_s (M_G; M_s = M_G) \right) \forall \ t \tag{7.11}$$

Equation (7.11) captures the essence of institution-building for altering owner-state investments. In this equation, the LHS corresponds to the net present value of the international subsidy when honouring the conservation commitment, while the RHS corresponds to the net present value of that same subsidy if the commitment is dishonoured and the subsidy invested elsewhere. The owner-state's decision problem concerns whether to invest the subsidy received in any given period back into the international regime (in hopes of receiving future similar subsidies), or whether to invest it elsewhere in the economy. This equilibrium condition states that the supplier state will invest in the international regime only to the extent that the global community assures a flow of future benefits (inclusive of subsidies) competitive with the return available from other productive assets (r).

As would be anticipated in decision-making regarding the state's portfolio, it is the future flows from portfolio decisions that matter, and these flows must all equilibriate at the prevailing marginal return on capital. Therefore, equation (7.11) demonstrates that the global community's interventionist role is the enhancement of the flow of benefits from diverse resources, so that they meet the competitive rate of return on other productive assets in the economy.

In addition, this intervention must provide the correct level of benefit enhancement *at each point in time*. The solution to the global problem of extinction requires institution-building in the sense that the alteration of investment decisions (i.e. portfolio selection) requires that intervention take the form of a flow. That is, the system of incentives required must be in effect at each and every point in time, because assets will attract investment in each period only if their return is perceived to be competitive under that period's conditions.

There is another important sense in which the resolution of this problem requires institution-building, i.e. the provision of assurance. Once the assumption of certainty is dropped (i.e. $0<v<1$), it is necessary to provide *and assure* an enhanced flow of benefits to states altering their investment portfolios. It is easy to see from eq. (7.10) that, if there is little credibility in the global community's offered flow of benefits (i.e. $v \to 0$), then there is no impact of the global community's intervention on the supply state's portfolio. The global community should invest resources in providing this assurance to the extent that they are warranted by the response of supply states (and hence the generation of global benefits).

In general, this would suggest that global investment in management should be distributed so as to maximise the likelihood of the supplier state observing its commitment, through investments directed both to increasing state benefits and to increasing assurance. Both an enhanced flow and the assurance of its availability in the future are equally necessary to generate changes in the supplier state's investment portfolio. Therefore, international investments in diversity management must be allocated between benefit enhancement (M_G) and benefit assurance ($v(M_G)$).

Optimal International Investments in Benefit Enhancement and Assurance:

$$v^*, M_G^* : \begin{array}{c} \text{Max} \\ M_G, v \end{array} \sum_s \left[\frac{v(M_G)}{1+r-v(M_G)} \cdot \prod_s (M_G; M_s = M_G) \right]$$

$$\longrightarrow \frac{1+r}{[(1+r)-v]^2} \sum_s \prod_s (\cdot) \frac{\partial v}{\partial M_G} + \frac{v}{(1+r-v)} \sum_s \frac{\partial \prod_s (\cdot)}{\partial M_G} = 0$$

$$\longrightarrow (1+r) \sum_s \prod_s (\cdot) \frac{\partial v}{\partial M_G} = (1+r-v)v \sum_s \frac{\partial \prod_s (\cdot)}{\partial M_G} \qquad (7.12)$$

Equation (7.12) represents the global objective which is to maximise the *expected value of future benefits* from the international regime. Expected benefits will be dependent upon the product of benefit level and likelihood, and the international community has the capacity to influence both. This indicates that the returns from investments in assurance should equilibrate with the returns from investments in state benefit enhancement. That is, the LHS of the last equation in (7.12) represents the increase in the perceived credibility of the global community in supplying future benefit flows, while the RHS represents the enhanced level of benefits. The global community must invest in both, because supplier state choice is determined by the product of benefits and credibility (i.e. expected benefits). There is no reason to invest in one without the other.

The method for providing assurance in contracting is termed institution-building. Here, institutions are forms of contracts which are intended to create credible commitments to a course of action over time (i.e. they create a higher degree of time-consistent commitment to a particular policy). This is done by undertaking investments which increase the costs of dishonouring subsidiary commitments in regard to these investments.

In private contracting, this is usually accomplished via 'bonding mechanisms', such as sunk costs (asset specific investments) which earn a return in only a single activity (Williamson, 1986). Credibility is enhanced when investments are asset-specific, because the opposite party recognises that reneging on the commitment involves a cost; flexibility is lost when assets are made expressly non-liquid. The additional costliness of these asset-specific investments renders a policy time-consistent over a wider range of possible conditions.

Contracts are of such a character when they involve substantial investments which are tailored specifically and committed irreversibly. An example of such a contract is a joint venture agreement that requires both parties to acquire assets that are useful solely in that venture. Such fixed overarching commitments make the other subsidiary commitments enforceable over a wider range of conditions.

Bonding mechanisms are equally applicable in public international law. They are useful whenever an external enforcement mechanism is either unavailable or costly. The global community is able to invest in the provision of assurance by using bonding mechanisms. That is, so long as the global community commits some assets irreversibly to a mechanism that promises an enhanced return to supply state investments in diverse resources (i.e. these assets of the global community are made 'specific' to the diverse resources), then the supply states are assured of the level of flow which these assets are able to generate.

This section has demonstrated the necessarily dynamic nature of the mechanism required to regulate extinction. It has also shown that this mechanism is likely to be of the nature of an 'institution', i.e. an independent agency with asset-specific investments of an indefinite duration. Independence and specificity are necessary to generate any level of assurance; indefiniteness is required in order to induce and enforce a commitment on the part of the supply states.

7.7 OPTIMAL DIRECT INDUCEMENTS: INTERNATIONAL FRANCHISE AGREEMENTS

An example of an optimal mechanism for regulating extinction of the first (direct) type would be an international agreement to transfer development rights. This is an often-mooted alternative for conservation, based on the observation that the global community frequently values the remaining wildernesses for other purposes (carbon storage, biodiversity) more than the owner-state does for development (see e.g. Schneider, 1992). This gives rise to the notion that some form of contract ought to be able to capture these 'gains from cooperation'.

Despite the differences in phrasing, the substance of the concepts discussed in these proposals is not distinguishable from that developed here. The difference lies in the analysis of commitment and enforceability. In an international transaction, the 'transfer of development rights' must be conceived of as a schedule of payments constructed so as to maintain the optimality of the globally-desired outcome. There is no other enforcement mechanism available to maintain the commitment of the owner-state overtime.

The decline of diverse resources within an owner-state can be remedied through a *system of strategic international payments*. This system of payments would necessarily be conditional upon the owner-state's application of them to the purpose of management of (and investment in) diverse resources. That is, the payments would be restricted to use for purchases of land and management for a particular resource of region. Of course, a crucial feature of any international scheme of payments based on conditionality (here, payment conditional on specific application) is its necessarily dynamic nature. That is, it is only possible to restructure the owner-state's decision-making process if the payment is offered at the end of each period that the state takes the specified action. The payment is the *global premium*, supplementing the state's return, received for pursuing the global rather than the local optimum.

An example of such an international factor-subsidy scheme might be an *international parks agreement*. Such a programme would in effect buy the use of the land for the diverse resources that exist there. The role of the international community would be to pay the rental price of the land each year that the national park remained unconverted. In order to manage the park, it would probably be necessary for the international community to provide a subsidy for management services as well. If the international community wishes to maintain diverse resources at the lowest levels of exploitation (e.g. non-consumptive activities such as tourism and filming), then it will be necessary to provide almost complete subsidies for the forgone land development opportunities and management services required.

However, the maintenance of a stock of diverse resources will often be compatible with a wide range of forms of resource utilisation, other than those that are of the lowest intensity. In that case the international subsidies may be reduced in accordance with the extent to which development opportunities are allowed in the region. This is, in essence, the *development rights* approach to diverse resource conservation; to the extent that the international community wishes to reduce the development intensity away from the local optimum (of high-intensity conversion and use), then it must be willing to provide a stream of ex-post payments to compensate for the forgone development.

Another example of an international subsidy scheme would be a *resource franchise* agreement. This would be a three-way agreement (between owner-state, international community and franchisee) in which the international community would provide a stream of rental payments to the owner-state in return for its agreement to offer a piece of land for use only for restricted purposes specified within the franchise agreement. That is, the land would be designated for use, but only for limited uses amounting to much less than complete conversion (e.g. extractive industries such as rubber-tapping and plant and wildlife harvesting). The land would then be franchised to an entity that is allowed to use it only for purposes compatible with the franchise. If the state fails to enforce the franchise agreement, it forfeits its year's rental fee.

The benefit of a franchise agreement over an international parks scheme is that it provides a stream of benefits to fund the management of the operation. Under the franchise agreement, it is the responsibility of the franchisee to provide management services from the returns it generates in its operation of the franchise. In this fashion, the international community is able to 'contract out' the provision of management services, while (through the rental-fee and restricted-use clauses in the franchise agreement) it retains the possibility of moving the local optimum closer to the global.

Therefore, a franchise agreement is simply a more generalised form of the international park agreement required for the acquisition of all development rights in a region. A franchise agreement could be used to specify the precise range of activities allowed and disallowed in the region, and it would be required of the international community to provide the 'global premium' that is necessary to make up the difference between what an entity would bid for the franchise and what is the value of the land in unrestricted use. That is, an international franchise agreement allows for the division of the 'development rights' pertaining to a particular region between the global and the local communities.

7.8   CONCLUSION

The problem of global extinctions is the problem of constructing the globally optimal portfolio of biological assets. This implies the necessity of international regulation in a sphere that to date has been managed entirely by individual states under the doctrine of national sovereignty.

International intervention in national resource management is complicated in the case of biological diversity by three factors. First, on account of national sovereignty over terrestrial resources, international intervention must be directed to the alteration of the decision-making frameworks of the supplier states, so that they themselves elect to construct an asset portfolio that is compatible with the global optimum. Specifically, the problem is to enhance the rates of return on internal investments in diverse resources so that the state invests in a different portfolio of assets. Second, the nature of the intervention is necessarily a long-term one, because it is an intervention to alter basic investment decisions. This also means that the mechanism for intervention must be both flexible enough to adapt to changing conditions, and stable enough to induce investments. Finally, the core problem of biodiversity lies in the intangibility of many of the goods and services flowing from diverse resources. This makes it very difficult to generate a compensating flow of funds in return for these goods and services. Equally, it makes it difficult to generate international support for the long-term investments that are required for the institution of a long-term solution to this global problem.

This chapter has outlined the nature of the solution to one-half of this problem, i.e. the manner in which national portfolio selections might be altered. In short, it has advocated *institution-building* capable of enhancing and assuring the benefits from diverse resource management. It has defined the three general forms that this new institution-building might take. This first (discussed here) is centred on direct subsidies to the supplier-states for low-intensity utilisation (franchise agreements). The second (discussed in Chapter 8) is centred in the consumer-states, and offers enhanced benefits from the trade in the tangible flows from diverse resources (wildlife trade controls). The third (discussed in Chapter 9) is the possibility of rendering the non-appropriable flows from diverse resources appropriable (intellectual property rights).

Institution-building is the required solution in all three cases, of the nature of the development of a mechanism for continuously transferring global values (of diverse resources) into the decision-making environment of local users of these same resources. Although the channel developed by each of the three instruments is distinct, the object is the same. The institution to be built must be able to register the global community's preferences for the low-intensity utilisation of the remaining wildernesses of the world. This requires the creation of a system of payments for the owner-states' rights to more intensive development. The general object, with regard to international institution-building for biodiversity conservation, is the development of institutions that create diverse pathways to development.

8 International Regulation of the Wildlife Trade: Rent Appropriation and Biodiversity Conservation

8.1 INTERNATIONAL INTERVENTION AND THE 'EXTERNAL' REGULATION OF DOMESTIC RESOURCES

Ultimately, state investments into the management of diverse resources will be constrained by the benefits that such investments are able to generate. International intervention that is based upon the direct induce-ment of further state investments in diverse resources will be similarly constrained. However, there are instances in which *external regulation* of the resource (i.e. in its product markets rather than its place of production) will yield greater returns per unit of investment. In that case, it may be optimal for the global community to utilise the instrument of external market regulation, generating supplier-state benefits through such intervention in consumer-state markets.

The regulation of the trade in the tangible goods of diverse resources (the 'wildlife trade') is a case in point. Although these resources are sourced, by definition, in the world's last refuges of diversity, the con-sumers of these products are often situated in the states with little diversity remaining. Turning once again to the African elephant for illustrative purposes, although the thirty host states reside within sub-Saharan African, the vast majority (about eighty per cent) of the tangible products from the African elephant have found their way to final consumers in the North: the USA, Europe and Japan (Barbier, *et al.*, 1990). Since the flow of the wildlife trade through this 'pipeline' from South to North is fairly common for most diverse resources, this affords the possibility for 'Northern'-based regulation of this exclusively 'Southern' resource. When it is both more cost-effective and rent-productive to base regulation in the North, then an 'external method of intervention' is the instrument of choice.

All of the necessary conditions for the construction of an effective international agreement apply equally here. In particular, this manner of global intervention must remain dynamically consistent through conditionality on state management spending. That is, as depicted in equation (7.8) above, the global community must invest each period to maintain the *external regulation mechanism*, in response to the supply state's management spending. This external mechanism must then automatically confer the enhanced benefits upon the investing supplier states.

However, there is another role that an external regulatory mechanism must fulfill. The creation of enhanced benefits from trading the tangible goods and services of natural habitats will not necessarily generate increased investments in those habitats. This is because the enhanced benefits of wildlife trade will also generate increased pressure on the natural habitats, and thus increase the costliness of managing natural habitat production. The rents from diverse resource flows are equal to the gap between revenues received and the costs incurred per unit of production, and it is very likely that both revenues and costs will move together in the context of natural habitat management.

A constructive wildlife trade regime must be developed in order to fulfill this dual role: the creation of incentives for investment (through rent maximisation) and the minimisation of the costliness of diverse resource management. To some extent, these two objectives are in conflict; however, the optimal institution for regulating wildlife trade will take both into consideration.

This chapter considers the institutional detail of the solution to this problem. It is the most detailed chapter, institutionally, precisely because there are international institutions already existing within this area. These are the institutions put into place under the Convention on International Trade in Endangered Species (CITES). This chapter constitutes a comparative case study of both the nature of the institution that would satisfy the requirements of a constructive trade control mechanism and the institution that currently exists, i.e. CITES. It demonstrates the practical and institutional difficulties that underlie the creation and implementation of a regime for the transfer of enhanced benefits from North to South.

8.2 THE INTERNATIONAL REGULATION OF WILDLIFE UTILISATION – AN ECONOMIC FRAMEWORK

Contrary to popular perceptions, a strategy of wildlife utilisation often has much to commend itself as a conservationist tool. It generates revenues for conservation and (in some cases) it creates incentives for the application of these revenues to that purpose. The latter objective is equally as important as the first. This has become apparent with the large number of conservation failures, such as the establishment of 'paper parks' (i.e. established, but unmanaged national parks). It is the fact that it has the potential to combine these capacities, i.e. fund raising and incentives generating, that makes a strategy of wildlife utilisation so appealing.

However, it is also apparent that a simple strategy of unregulated wildlife utilisation is not capable of generating these incentives in and of itself. Thus, wildlife exploitation remains a significant threat to many species. This happens whenever wildlife utilisation generates revenues but not rents. The difference is crucial, as revenues are simply returns that result from the labour expended in an activity (such as the hunting of wildlife) while rents result from the creation of a return to resource ownership. Where there are revenues to be earned from resource utilisation, but not rents, the incentives are to exploit but not conserve the resource.

Therefore, the object of a constructive *wildlife trade regulation mechanism* (in the sense of the creation of incentives to wildlife conservation) must be the creation of *rents* in the wildlife resource. What are the predictable difficulties that arise in the attempt at installing such a system? This chapter considers both sides to the problem (demand and supply) and discusses the nature of the international environmental agreement required to address them.

First, the 'demand-side' problems of wildlife utilisation will be considered. Here, the primary limitation of a utilisation strategy is its inapplicability to species which have low 'consumptive' use values. That is, some species happen to have developed prior usefulness and thus have a means of 'paying their way'; others, however, have not. It is problematic that a strategy of wildlife utilisation extends expressly only to the conservation of the former and not to the latter.

The greater part of this chapter will focus, however, on the 'supply-side' difficulties with wildlife utilisation. This is because it is often the case that the management of natural habitat utilisation is both difficult and costly to control, and this is one of the primary reasons why overexploitation has so often occurred in the context of natural habitats. Some of these difficulties are relatively easily remedied, for example by devolving the habitat from state to local community control. (Swanson and Barbier, 1992). However, these problems can remain costly nonetheless, on account of the less intensive techniques used for control (e.g. lack of fences). And, in some instances, the conversion of natural habitats to more intensive forms of production (e.g. ranching or farming) will occur simply on account of these cost considerations.

A 'constructive' strategy of wildlife utilisation must take these two factors into consideration, and develop mechanisms that address both. The demand-side considerations (concerning the non-consumptive values of certain species) can be addressed through rent-creation policies, involving the construction of a *quota system* or its equivalent. The supply-side considerations (concerning the costliness of natural-habitat control systems) can be addressed through selective purchasing policies, involving the invocation of *conditionality* in consumer purchasing systems. Together, an effective constructive wildlife regulation system is capable of creating rents and channelling them to the conservation of species (Swanson, 1992a).

8.3 DEMAND-SIDE MANAGEMENT OF TRADEABLE WILDLIFE RESOURCES

The point of demand-side management is simply the maximisation of the rents potentially appropriable by the owners of the resources. All species have value, and much of this value resides in the Northern developed countries. One facet of this value is consumptive in nature, i.e. individuals are willing to pay for the species' existence for the purpose of direct use of its characteristics. In order for the maximum number of any species to exist, the full extent of each type of value must be appropriated by its (and its habitat's) managers. This is the process of maximising the rents to the owners.

Revenue maximisation is a precondition to rent maximisation, and it is achieved through the control of the price at which the tangible products from natural habitats are sold. All goods have a certain price at which revenue maximisation occurs, and if there are not good substitutes for them, that price can often be quite a bit higher than that which exists in the market. The fact that natural habitat is becoming more scarce, and is expected to become even more so, contributes to the relative scarcity of its products, and this makes its revenue maximising prices (and the total revenues potentially appropriable from them) even higher (Krutilla, 1967). In short, one of the conseqences of centuries of conversions of natural habitat to more specialised production systems is to increase the inelasticity of demand for the flow of goods from the former (as supplies of these goods and their closest substitutes become more restricted) and to decrease the elasticities of the goods from the latter (as their supplies 'flood' the market) (Swanson, 1990a).

This is the basis for the expectation that the demand for the tangible products of increasingly scarce natural habitats will usually be inelastic. This was found to be the case in a recent study of the demand for African elephant ivory, for example, where it was found that an increase of 100 per cent in the price of ivory in Japan would result in a fall in quantities of only 70 per cent, thereby resulting in an overall increase of 30 per cent in the total revenues flowing to its suppliers (Barbier, *et al.* 1990). Therefore, there is both a theoretical and an empirical basis for the expectation that revenues from the sale of the products of natural habitats might be enhanced through carefully-orchestrated price controls.

The maximisation of the revenues from the consumptive use of wildlife resources is important precisely because it is representative of so much value that is *non-consumptive* in nature. That is, the regulation of the wildlife trade should be considered to be simply the regulation of those few bits and pieces of biological diversity that are valued on the market.

The maximisation of wildlife utilisation revenues provides the means of indirectly compensating for these inappropriable values. When it is clear that there are important positive, inappropriable, external values closely associated with the provision of a tangible good, there is good reason for society to increase the reward to the tangible good in order to provide an implicit return to the intangible one. That is, when these products come jointly (tangible/intangible) but it is only possible to price one of the two goods (because the other is of the nature of a 'public good'), then it can be the optimal 'second-best' policy to provide price supports to the tangible good in order to foster the incentives to invest sufficiently in the habitat which supplies both.

In essence, in this framework, all of the species inhabiting a given parcel of natural habitat are being viewed jointly. While all of these species have 'value' (in the sense that we would like to see them continue to exist), some of them have 'consumptive values' and others do not. Since it is very hard to require payment of 'existence values' to the owners of the natural habitat, an alternative means of acquiring additional value is to instead increase the amount of the 'consumptive value' appropriated. To some extent, the surplus consumptive values captured by the landowners with respect to a few species replaces the inappropriable non-consumptive values of many other species, and restore the correct incentives for investing in unaltered habitat. The parcel of natural habitat is kept for the use of all of its resident species, and paid for by the price supports on the consumption of a very few of them. This is the rationale for encouraging the creation of mechanisms for rent appropriation in wildlife utilisation.

8.3.1 The Nature of Rent-Appropriation Mechanisms

One of the most substantial obstacles to increased investments in diverse resources is the generally low rate of resource rent appropriation in the supply states. That is, very frequently, diverse resources are sold in product markets at or near the cost of harvesting them (labour costs, capital costs); there is often no value received in these markets that is attributable to the natural resources itself (Swanson, 1991b).

It is clear that an open-access management system will produce this result. With no coordination of the harvesting sector, the harvesters (if there are enough of them) will harvest until price is driven down to the costs of access, i.e. zero natural resource rent (Gordon, 1954; Clark, 1976). If there were sufficient returns available to the management of this commons, it is predictable that the state would invest in altering the prevailing management regime; this is demonstrable from the state's decision problem set out in equation (5.15). However, it is very often the case that the supply state does not invest in its diverse resources.

One reason for this is that there are in fact two 'layers' of commons involved in the production and marketing of the resource, and the individual supplier-state only has control over one. There are limited returns available to expenditures on management when they only reach a part of the management problem.

The two layers of the natural-resource commons can be demonstrated in a simple example. Consider first a state with a unique natural resource that is under first-best management ('full coordination'). The perceived market value of the products of this resource will be managed to achieve maximum unit value (which will then be equilibriated with the rental value of the resource).

Rental Value – Full Coordination

$$Y^* : \begin{array}{c} \text{Max} \\ Y \end{array} p(Y) \cdot Y - c(x) \cdot Y$$

$$subject\ to: \dot{x} = H(x, R) - Y$$

$$\longrightarrow Y^* : p\left(1 + \frac{1}{\varepsilon_D}\right) - c = \lambda_G \tag{8.1}$$

This optimality condition states that the fully-coordinated state perceives the true price elasticity of further production. When further production results in revenue reductions, this condition is recognised and internalised in the management decision. Since the resource is 'unique' within this state, national first-best management will then correspond to global first-best management. The resource rental value prevailing under these circumstances is termed λ_G, corresponding to the maximum 'globally-managed' rental value of the diverse resource.

Now consider a slightly more realistic situation, where there are a number of different states (N) harbouring the same natural resource, although initially it will be assumed that each of these states invests to achieve perfect internal coordination. In that case, each of these N states will be faced with the following production problem (solved with use of simple Nash assumptions).

Rental Value – N States, Internally Coordinated:

$$y^* : \begin{array}{c} \text{Max} \\ y \end{array} \quad p(Y) \cdot y - c(x) \cdot y$$

$$subject\ to : \dot{x} = H(x, R) - y$$

$$\longrightarrow \frac{\partial p}{\partial y} \cdot \frac{\partial Y}{\partial y} \cdot y + p(Y) - c(x) = \lambda_S$$

$$\longrightarrow y^* : p(y)\left(1 + \frac{1}{\varepsilon_D \cdot N}\right) - c = \lambda_S \tag{8.2}$$

Under Nash assumptions, each of the N states assumes that the other supplier-states will not respond to its expanded production. In that case each state has the incentive to continue expanding production beyond the point that is jointly optimal, because each fails to recognise that the others have the same incentives. In the end, each state is only recognising ($1/N$) of its full impact on joint revenues. The resulting unit rental value is termed λ_s, corresponding to the rental value deriving from the first-best national management in a context of uncoordinated global management.

Now consider a far more realistic scenario, with N producer-states each containing n potential producer individuals. The N producer-states are not coordinated in their marketing of the diverse resource. It is not assumed, however, that open-access conditions prevail internally; rather, each state invests in management to reduce the number of conflicting strategies in the internal commons to ($n - \mu$), the locally optimal outcome. Each of these individual harvesting units then makes the production decision (y_i).

Within the context of these two layers of management, what will be the resulting harvest of the resource and what will be its unit rental value?

Rental Value – N States, (n – μ) Individual Harvesters

$$y_i^* : \begin{array}{c} \text{Max} \\ y_i \end{array} p(Y) \cdot y_i - c(x) \cdot y_i$$

$$subject\ to: \dot{x} = H(x, R) - y$$

$$\longrightarrow \frac{\partial p}{\partial Y} \frac{\partial Y}{\partial y} \frac{\partial y}{\partial y_i} y_i + p(Y) - c = \frac{\partial y}{\partial y_i} \lambda_i$$

$$\longrightarrow y_i^* : p \left(1 + \frac{1}{\varepsilon_D N(n - \mu)} \right) - c = \lambda_i \tag{8.3}$$

Another layer of necessary management introduces another inefficient distortion into the exploitation framework. Individual producers fail to perceive their full impact on joint revenues both with regard to other domestic producers and also with regard to other extra-national producers. They are uncoordinated, both internally and externally. The prevailing unit rental value (in state *s*) is termed λ_i, corresponding to the rental value deriving from uncoordinated national and global management of the resource.

In practice, equation (8.3) probably represents the management incentives affecting almost all diverse resources. Individual states have an incentive to invest in some management services (thereby decreasing the amount of uncoordinated harvesting $n - \mu$) in order to increase the market value of diverse resources. However, the impact of this incentive is diluted by the unmanaged access of other states to the global market. Even if internal management is first-best ($n - \mu = 1$), there is still a distortion in the incentive framework on account of the presence of N producer states in the product market. Therefore, even if a diverse resource is potentially profitable, if production is properly managed, there may still be a disincentive to investments in domestic management by reason of non-coordination with other producer states.

This indicates the nature of the second route for intervention by the global community. In short, recall that the driving motivation for state management spending is to increase state benefits through increased unit resource rents (as in equation (7.6)). Equation (8.3) demonstrates that the unit resource rent available within an individual state might be more easily regulated externally than internally (i.e. $d\lambda/dM_G > d\lambda/dM_s$). Since the ultimate objective of global intervention is to increase the state benefits received from diverse resources (compare equations 7.6–7.8), external regulation should be used to do this when it is more cost-effective than internal.

The precise form that this manner of international regulation takes mirrors the state management of the domestic commons, analysed in Chapter 5. The objective is to invest sufficient resources to remove some producer-states from the international market place. That is, the object is to remove M producer-states from the market, leaving the market to the remaining $(N - M)$. In that case, the actual operation of this method of intervention is as follows (a revision of eq. 7.8)

Rent Appropriation Mechanism (as an inducement to M_s)

$$M_G^{l_2} : \frac{\partial \prod_s}{\partial \lambda_s} \frac{\partial \lambda_s}{\partial (N - M)} \frac{\partial (N - M)}{\partial M_G^{l_2}} \frac{\partial M_G^{l_2}}{\partial M_s} \tag{8.4}$$

Here, management spending is being induced from the supplier state by means of a commitment to restrict the number of operators in the international market, and these restriction then generate the enhanced unit rents that benefit the supply state. In essence, the global community's investment is directed to the replacement of λ_i with λ_G for states investing in their diverse resources. Such a mechanism creates an incentive to invest (in order to get into the mechanism) and then sustains that incentive (through increased benefits from internal management expenditures).

In principle, it is possible for the global community to generate the fully-coordinated resource rental value, as there is no reason why there need be more than one 'supplier' to the international market.

All that is required is that the consumer states agree to make all purchases from a single point of supply, which then acts as a conduit for all supplier states. At this point in the market, coordination of all states' production occurs in order to maximise aggregate rents. In essence, then, a market unit rental value of λ_G is available to each of the states supplying this conduit, and the incentives for investing in the resource are based around this value.

8.4   SUPPLY-SIDE MANAGEMENT OF WILDLIFE RESOURCES

The international community has the ability to use its power (as the consumer of wildlife products in international trade) to create rents in wildlife products. Difficult problems are created simultaneously, however, because the creation of rents will simultaneously create a group of opportunists who will seek to capture these rents, to the detriment of the range states and the wildlife resources.

Therefore it is necessary for the international community to create a system for the careful discrimination between different groups of suppliers of the wildlife products; otherwise, it is predictable that a rent-creation strategy must fail. The cost of producing wildlife 'sustainably' will always be greater than the cost of open-access exploitation (because the former includes an implicit rent to the resource). Without an effective mechanism for discriminating between the two types of suppliers, the unsustainable producers will drive the sustainable from the market (Akerlof, 1970).

The remainder of this chapter is an analysis of the necessary components of an international regime for the discrimination between producers, and a comparison of the existing regime to investigate how it performs as a constructive trade control mechanism.

8.4.1 Stock Conditionality

The international community wishes to confer the enhanced benefits from demand management only upon those states that generate global benefits from their diverse resources. This implies that there must be some conditionality on the assignment of enhanced benefits, and this conditionality must relate to owner-state investment in the maintenance and management of diverse resources. The requirement that diverse natural habitats be conserved for low-intensity uses will be termed *stock conditionality*. The essence of supply-side management is to create a system that effectively discriminates between the flow of wildlife goods from investors in diverse resource stocks and the flows of such goods from all others.

Certification is the process that enforces stock conditionality. Suppliers would be certified on the basis of their demonstrated management of the natural habitat and wildlife resources for which they are responsible. Specific criteria indicative of the sustainable management of natural habitat (regarding the nature and extent of development in the species' habitat) would need to be included within the trade control regime in order to provide the basis for differentiation between 'certifiable' and 'non-certifiable' suppliers.

This system of indicators must be applied periodically to determine whether a state is to be given 'certified' status, and also to determine whether that status is to be retained. In this way, consumer state purchasing is made conditional on diverse resource stocks in each and every period.

Such a tailored control regime fosters an international system of incentives toward sustainability, as potential suppliers pursue certified status and the already-certified strive to retain it.

To this point, there is little to differentiate the analysis of this chapter from that of Chapter 7, other than the attachment of the global premium to the wildlife product. The remainder of this chapter, however, examines the particular costliness of this manner of benefit enhancement, i.e. the cost-liness that comes from attempting to segregate between products that have been 'sustainably' and 'non-sustainably' produced.

8.4.2 An International Mechanism for Controlling Access to National Commons

A primary objective of an effective supply-side policy is to reduce the costliness (or to increase the cost-effectiveness) of owner-state manage-ment of designated natural habitats. This is because there are always two factors placing pressure on natural habitats: the costliness of its manage-ment in low-intensity uses and the opportunity costs of its development for high-intensity uses. In addition, attempts to combat the latter (via a wild life rent creation strategy) will consequently increase the costliness of the former (by increasing pressure on the habitat). Therefore, the effectiveness of an international regime to create rents depends upon the capacity of that regime to reduce the costliness of managing the pressures that it creates internally.

This problem is complicated by the fact that it is not desirable to move toward simple, intensive management of the habitat. In general, it is contradictory to speak of intensive controls in connection with natural habitat. One of the primary attributes of such habitat is its unregimented nature; this is what allows diversity to flourish. Therefore, one of the requisites of an effective supply-side policy will be its capacity to control supplies without intervening drastically within the habitat (by the use of fences, massive patrolling, etc).

A policy of selective purchasing solves the problem of management costliness in theory. A consumer-enforced commitment to purchase only from the designated supplier accomplishes this. In short, there is no incentive to engage in unauthorised harvesting if there is no market for the produce.

However, there are always 'black markets' which undermine such con-trols. The extent of an illegal market is largely determined by the method of implementation and the consistency between the objectives of the controller and those being controlled. With regard to the latter, the vast majority of the *ultimate* consumers of many wildlife products reside in developed countries, where there is a widely-expressed demand for more effective controls over natural habitat production. These persons can be expected to support a more effective system of trade controls.

This has been demonstrated in the case of the recent ivory trade controls implemented by the primary consumer states (the EC, the US and Japan), where the end consumers have effectively shut down the trade by refusing to purchase ivory products. Prices of ivory products have fallen precipitously in the US and EC end markets, while remaining fairly stable in Asia (Luxmoore *et al.*, 1989).

Therefore, unlike consumer state controls on the import of substances widely demanded by the citizenry of those states (e.g. various narcotic substances), the objectives of the end consumers of wildlife products are often fully consistent with the reduced flows of those products. When this is the case, it is possible to implement low-cost regulation based upon selective purchasing agreements.

Even assuming that the objectives of wildlife producers (the investors in diverse resource stocks) and wildlife consumers (the consumers of the end products) are broadly consistent, there remain other, intermediate groups whose interests are conflicting. An international trade mechanism must still deal with the problems raised by entities that come in between a direct producer–consumer exchange.

8.4.3 The Problem of Cross-Border Incursions

Even with a 'designated supplier' system, there still remain incentives for the excessive exploitation of natural habitat. This is because each supplier has no incentive to restrict flows derived from natural habitats other than its own. In essence, conditionality which is based upon the status of a states's own diverse resource stocks provides no incentives to care for the resources of other states. Therefore, a system of conditionality with a 'narrow' focus (on only the stocks of own-state resources) can encourage cross-border incursions as suppliers are willing to trade in any wildlife that does not visibly affect its own stocks.

The solution to this dilemma is for consumers to adopt *flow conditionality* as well. This would restrict the purchasing of quantities from a designated supplier to an amount that would correspond approximately to the quantities which could feasibly issue from the sustainable management of that supplier's own diverse resource stocks.

The conditionality of certification must then be based on both the state of the diverse resource stocks *and* the maximum sustainable flows from those stocks. A consumer-enforced system of individual owner-state quotas is the result of the second requirement. Given individual quotas, the owner-state has absolutely no incentive to look elsewhere for supplies, as the correct management of its own diverse resource stocks will satisfy its quota (and correct management is a prerequisite to the retention of its certified status).

In addition, given that the same system is applied across all other certified owner-states, there is no longer any incentive for any of them to pursue others' supplies. Each supplier's habitat is secured because the size of its market is secured; cross-boundary excursions cannot increase the size of that market and so there is no longer any incentive to undertake them.

Such a system could effectively be implemented through a number of different mechanisms. Individual quotas – based on the anticipated sustainable flow of products from the correct management of designated natural habitat – could be established by a technical committee. These quotas could then be enforced through the issuance of 'coupons' or the construction of an 'exchange'. Coupons essentially take the place of the regulated commodity in the market. Once a given number are printed, then the consumer states enforce the regulatory system by refusing to admit the regulated good without a coupon. When the coupons are distributed to the designated suppliers, the task of controlling access to the species becomes one of controlling access to the coupons instead, which is usually a far less costly proposition (Swanson, 1989b).

Alternatively, the consumer-states could instead construct an 'exchange', which is a mechanism for assuring that goods on the market are 'exchanged' only between consumers and designated suppliers. An exchange operates by allowing certified trade to occur only within its walls, while consumers enforce the exchange mechanism by refusing entrance to any goods other than those traded through the exchange. Again, the exchange would have to establish and enforce individual quotas for each designated supplier, disallowing sales within the exchange beyond the annual quota (Swanson and Pearce, 1989).

8.4.4   The Problem of Manufacturer States

Both of the above options hinge upon the satisfaction of one of two conditions for their effective implementation: *either* (1) all existing and potential consumer-states agree to enforce the quota (and this implies the necessary full cooperation of all states in the world as, for the consumer state above, it should be assumed that this includes all consumers of the worked and the *unworked* natural product, and the latter has been demonstrably mobile in the past): *or* (2) all final consumers must agree to enforce the quota, and there is no allowed re-export of worked or unworked natural product (which implies the necessary full cooperation of only the primary consumers: the US, the EC and Japan).

The first of these options is a pipedream, and its pursuit has brought international trade regulation into ill-repute. It cannot work for two reasons.

First, any given state in the world can be a possible manufacturing state. That is, although the raw material and the final consumers are relatively fixed, the labour necessary to work raw materials is very fluid. Therefore, a regime based on regulating the manufacturers (rather than the final consumers) of natural habitat materials must necessarily have the active cooperation of all the states of the world. This is not possible. There are too many states, and there is also a positive return to be had from 'breaking' the system. Controls based on this mode of operation will only redirect, not restrict, the trade.

Secondly, although the incentive to 'free-ride' exists (in favour of deviating from the system) irrespective of whether the state is a manufacturer or a consumer of the raw material, there is little countervailing incentive in favour of conservation in the manufacturing state. In the consumer-state, however, the raw material is valued for its final use, and there will be an interest in its long-term conservation. For the manufacturer, raw material is more of an object with which to mix labour prior to sale; usually, there is a good prospect for moving to another, different material if the first comes into short supply. The manufacturer is, in essence, selling its skills more than the raw materials (for which it is paying the full value to the producer-states).

Therefore, in most instances, only the latter option exists, i.e. restricting all re-exporting of the natural materials. Initially, this sounds more harsh than it is in actuality, as its only implication is that the manufacturing sector must move its facilities to the consumer markets and import the raw material there for working and final sale. Given the demonstrated fluidity of most of these manufacturing industries (in both wildlife trade and other industries), this should not ordinarily be a major problem if the manufacturer is truly interested in the commodity.

8.4.5 Layers of Selective Purchasing Programmes

In the event that a problem with the relocation of the manufacturing sector does arise, there is an alternative, but it is generally too costly and cumbersome to be effective. This alternative allows the manufacturing sector to remain within one state while exporting to another, but only on the condition of consumer state enforcement of the control system as against the manufacturing states. Again the system hinges on consumers', rather than manufacturers', support – but this time the consumers' threat to withdraw custom applies to non-viable manufacturers, not to the producers of the raw materials. It requires the manufacturers to conform to the selective purchasing system (with regard to raw material purchasers) through a final consumer-based system of selective purchasing that is applied to the manufacturers.

This could operate through a 'restricted marking' system, whereby each final product for sale in a consumer-state is provided with a numbered mark corresponding to the licence or permit under which it was imported into the manufacturing state. It would fall to the consumer-states to maintain full records on their own imports of manufactured materials as well as the manufacturers' imports of raw materials, *and* then to perform the complicated task of matching up manufactured quantities with licensed imports.

This is a much more complex and costly system, and so its costliness will often dictate a policy of prohibited re-export instead. However, it demonstrates that the conservationist interest of final consumers can always be brought into use in trade regulation, but sometimes in a multiple-layered regime.

8.5   DEMAND- AND SUPPLY-SIDE MANAGEMENT COMBINED

Therefore, just as there are a number of demand-side options for controlling the international trade in wildlife products, there are a number of supply-side options as well. For each 'industry', constituted of range-states/manufacturers/consumers, there will be a first-best system of controls which will combine some manner of demand-side controls with the best supply-side controls for that situation.

The combined system has as its objective the creation of *rents* in the wildlife resource and the channelling of those rents to the local managers through a system of *conditionality*. The key to the entire system is the enforceability of this conditionality in the context of natural habitat production. The system must create rents for the flows from diverse resource stocks, by focusing on the simultaneous support of prices and reduction of management costliness (but only for the certified producers).

The system commences with the commitment of funds by all of the *consumer*-states to the system's enforcement of the conditionality of purchasing. These funds are the means by which rents are created and allocated, and hence they generate the enhanced benefits from diverse resource management.

The system is only as effective, in general, as is the weakest significant consumer's enforcement efforts. For this reason, the restriction of the necessary consumers to the smallest number possible, and to the group whose objectives are most consistent with diverse resource conservation, is important. Diverse resource conservation through wildlife utilisation is very likely an objective shared by producer-states and end consumers. The regulatory problem often comes down to the elimination of intermediaries, whose interests are not similarly aligned.

One means by which the system can be made operable is through the elimination of intermediaries. A restriction on re-exporting accomplishes this. This creates the possibility for a direct purchasing agreement between owner-state and final consumer.

Therefore, an optimal wildlife utilisation regulatory system should move toward consumer-enforced 'commodity cartels', where the consumer-states act to enforce the price supports adopted by the producers' association *on the condition that production occurs on a 'sustainable' basis*. This will increase the returns to particular consumptive goods of natural habitat (and implicitly compensate investments in the supply of related non-consumptive goods as well); in addition, it will decrease the costliness of natural habitat production.

8.5.1 An Example of a Rent-Appropriation Mechanism – Wildlife Resource Exchanges

An example of this sort of mechanism would be a 'resource exchange'. Such an exchange would be the agreed supplier for all consumer states (group North), and the Northern states would then maintain the mechanism through continuous monitoring (expenditures) to prevent imports other than from the exchange. The exchange would purchase its supplies in an amount maximising aggregate rents. The aggregate quota would be allocated to participating states, and each would receive an enhanced unit rental value.

In consideration of the continued support by the Northern states for the exchange mechanism, the supplier-states would allocate the exchange quota to sustainable producers of the resource, i.e. states investing in the management of their resource stocks and habitat. These producers would then receive an exchange-determined resource rent value (λ_G), while other supplier-states would receive an effective price near zero (depending upon the level of Northern funding invested in supporting the exchange mechanism). This would create incentives for other states to invest in diverse resources in order to join the exchange.

To the extent that an exchange requires start-up costs, such as physical site and equipment, these would represent the asset-specific sunk costs required to institutionalise the mechanism. Such institutionalisation creates the assurance that is necessary for international regulation.

8.6   CITES AS A CONSTRUCTIVE REGULATORY MECHANISM FOR THE WILDLIFE TRADE

The only issue pursued in this section is the compatibility of CITES with the concepts developed in the previous section. That is, the questions addressed here are: Does CITES provide a constructive trade control regime, as defined above? That is, is CITES capable of generating rents in wildlife resources and directing these rents to wildlife conservation?

The CITES convention was signed in March 1972 and came into force three years later. It provides the primary international control structure for the trade in wildlife products. It is apparent that CITES was drafted with little attention to the concept of constructive utilisation of the trade. It focused instead on the identification of endangered species and the withdrawal of the demand for them. Recently, however, the Conference of the Parties has been taking steps toward a more constructive approach, with the attempted development of various sorts of quota systems. Although these are still in their formative stages, they represent important steps toward a rationalised international control structure.

8.6.1   The Control Structure within CITES

Of the large number of international environmental conventions, CITES has probably the single most detailed control structure. It was the first international wildlife treaty to provide for both express obligations and international monitoring. Therefore, CITES represents an important step along the road toward making substantive international law with concrete impacts. The purpose of this analysis is, however, to ascertain its consistency with an economic control structure – as outlined above. Other authors may be referred to for a detailed analysis of the other working of CITES (Lyster, 1986).

CITES functions as a potential control mechanism primarily through the operation of two Appendixes, on which potentially endangered species are listed. Appendix I is intended as a list of those species which are currently threatened with extinction [Art. II(1)], while Appendix II is to contain a list of species for which there is some indication that they might become threatened [Art. II(2)]. The Conference of the Parties to CITES makes these determinations at its biennial meetings.

Once a species is listed on either of the CITES Appendixes, it becomes subject to the permit requirements of the convention. An Appendix I species may not be shipped in the absence of the issuance of an 'export permit' by the exporting state [Art. III(2)]. And this permit may not be issued, under the terms of the convention, unless both the exporting state certifies that the export will not be detrimental to the species and the importing state certifies (by the issuance of an import permit) that the import will not be used for commercial purposes [Art. III(3)(c)]. Therefore, an Appendix I listing acts as an effective 'ban' on the trade in those species and, even if exporters wish to continue the trade, the importing states have the duty to deny all commercial imports.

An Appendix II listing, on the other hand, leaves the decision on trade control wholly to the discretion of the exporting state. That is, there is no role for the importing state, other than to ensure that an import permit is issued for each specimen [Art. IV(4)]. And these permits are allowed to be issued so long as the exporting state itself certifies that the export will not be detrimental to the survival of the species within the exporting state [Art. IV(2)].

The other important responsibility of member states is to provide annual reports to the CITES Secretariat on the amounts of trade in listed species [Art. VIII(7)]. The Secretariat also sometimes acts as the intermediary between exporting and importing states, in order to confirm the authenticity of trade documents, for example.

8.6.2 The Convention as a Trade Regulation Mechanism

As drafted, the CITES convention provides little in the way of a 'constructive trade control mechanism'. The history of the CITES convention has witnessed many species progress from Appendix II to Appendix I, as potentially unsustainable trade levels raise concerns about the viability of the species. Most recently, this has occurred in the well-publicised case of the African elephant, for which a 12-year listing on Appendix II ended in 1989 with its 'uplisting' by the Conference of the Parties.

Such a progression from 'potentially threatened' to 'endangered' is predictable, given the structure of the CITES convention. This is because an Appendix II listing gives little in the way of a wildlife trade control framework; that is, it does nothing to manage either the supply or the demand pressures in the manner set forth above.

In regard to supply management, it leaves each range-state operating independently, with no international assistance to perform the additional tasks that are required of the parties. Therefore, an Appendix II listing provides only additional tasks, and no real incentive framework, for the control of the trade supplies of listed species.

With regard to demand management, the impact of CITES could range quite widely. It is possible that an Appendix II listing might result in individual consumers withdrawing their demand for the listed species; however, this is in general the current function of an Appendix I listing. What an Appendix II listing certainly does accomplish is to publicise the potential rarity of the listed species. For some species this might result in an increase in the speculative demand pressures. Therefore, the demand-side impact of an Appendix II listing is uncertain, and it could range quite widely.

Where Appendix II listing does encourage speculative demand, the combination of additional demand pressures with negligible control structures is a threatening one for Appendix II listed species. Since most wildlife species exist in unmanaged circumstances, it is usually difficult to handle the existing, let alone any increased, pressures. Thus, the progression of species from listing on Appendix II to 'endangered' status is not unforeseeable; for some species, it is entirely predictable.

An Appendix I listing promises much more in the way of international cooperation; however, the efforts are put to no constructive effect. That is, once the regulated species completes the progression from virtually uncontrolled Appendix II species to endangered Appendix I species, the international community then launches into concerted action to 'ban the trade'.

Of course, this will address the problem of possible immediate extinction from overexploitation which might arise during an Appendix II listing; however, it does nothing to provide resources for the management of the species in order to avoid extinction in the medium term. For the vast majority of endangered species, the withdrawal of value necessarily hastens the process of its elimination. They are under threat from habitat conversion. To address the force for conversion, in anything other than short-term circumstances, requires management and finances.

In the worst cases, the situation has gone from bad to worse; with the uplisting to Appendix I, the traded species is protected from short-term extinction from overexploitation, while simultaneously hastening the medium-term extinction of its habitat. Now what is lost is not only the single species generating consumptive value, but also many of the related species and systems whose shared habitat could be subsidised by this value.

Therefore, CITES as drafted provides for a peculiar sort of international regime of trade controls. For traded wildlife species it initially provides virtually nothing in the way of an international control structure, together with a global notice of 'potential rarity value' (with the posting of an Appendix II listing), while following that with the withdrawal of developed-world purchases of the wildlife product (via an Appendix I ban) if the population then comes under even greater pressures. The former, at best, provides no positive incentive framework; the latter provides no possible constructive use of wildlife consumptive value.

As drafted (i.e. considering only the Appendix I/Appendix II options), CITES provides for little in the way of a 'constructive trade control mechanism'. In essence, this means that it does not constructively address the two forces threatening most extinctions, and thus it does not adequately provide for their control. The direction for the reform of CITES – as a constructive trade control mechanism – will be to create a regime which simultaneously addresses both causes of extinction: over-exploitation and conversion.

8.6.3   The Convention – Recent Attempts at Innovation

As early as 1979, the delegates from developing countries brought the anomaly of 'indirect extinction in lieu of direct overexploitation' to the attention of the Conference of the Parties. In San José, Costa Rica, they argued that there must be an economic benefit from the controlled species if they were to be able to justify protecting their habitats from development. These concerns gave rise to the first step towards the reform of CITES, with the adoption of Conference Resolution 3.15 at the New Delhi Conference of the Parties in 1981. This resolution provides for the down-listing of certain Appendix I populations for the purposes of sustainable resource management. The criteria which specify how Appendix I species may be utilised in order to procure compensation for their habitat are known as the 'ranching criteria', and each Conference of the Parties usually sees a large number of such proposals for review and possible acceptance. The first ranching proposal accepted involved the transfer of the Zimbabwean population of Nile crocodile to Appendix II in 1983 (Wijnstekers, 1988).

Ranching proposals tend to be focused on a particular state, or operation, and do not constitute mechanisms for the constructive control of the entire trade. In essence, they continue the 'ban' in effect while allowing very limited, individual operations to recommence. While being of some utility, they do not constitute attempts at harnessing the value of an entire species for its own conservation.

In 1983, a species-based approach was first adopted with regard to the exploitation of the African leopard. Although listed on Appendix I, it was recognised in Conference Resolution 4.13 that specimens of the leopard could be killed 'to enhance the survival of the species'. With this, the Conference of the Parties approved an annual quota of 460 specimens, and allocated these between the range-states. In 1985 this quota was then increased to 1140 animals, and in 1987 to 1830.

This approach to trade management was then generalised in 1985 with Resolution 5.21, which provided for the systematic downlisting of populations where the countries of origin agree a quota system which is sufficiently safe so as to not endanger the species. Under this Resolution, five different species have been subject to quota systems: three African crocodiles, one Asian crocodile, and the Asian Boneytongue for which the Indonesians were allowed a quota of 1250 specimens (the latter being a fish much admired by the Japanese as a wall hanging).

None of these ranching systems went any further than the development of species-based quotas. In particular, no external control structure was ever implemented, this being left to the discretion of producer-states. Thus, predictably, these quotas can be abused. For example, Indonesia was reported to have issued permits for about 140 per cent of its first year's quota of Bonytongues.

The third avenue of innovation under CITES, and the most concentrated effort thus far at the development of an international control structure within the system, was the creation of a Management Quota System for the African elephant populations under Resolution 5.12. This system was founded upon the idea of management-based controls with consumer-based enforcement. Annual quotas were to be constructed at the outset of each year, and producer-states were then to issue permits not exceeding these quotas. Then consumer states were to disallow all imports unless accompanied by a Management Quota System permit.

At first glance, this may sound very similar to the constructive trade control regime outlined above, but there were several fatal differences (although the conceptual basis was similar). These differences indicate the crucial importance of the detail in an effective control system. First, the Management Quota System provided no external checks on the discretion of the producer states. The determination of annual quotas and the issuance of MQS permits was within their unsupervised discretion. This resulted in most states basing their annual 'management quotas' of ivory on the 'expected' confiscations from poachers. That is there were no enforced incentives for sustainable use, and also no disincentives for cross-border exploitation. Burundi, with a single elephant within its borders, became the largest exporter of ivory in Africa under this control regime.

Secondly, the system relied on the good faith of all consumers, including the trading states, for its enforcement; predictably, the good faith of these persons was insufficient to manage demand pressures. Consumer-states attempting to enforce management systems were unable to distinguish between properly-issued and improperly-issued permits. Many of the permits were patently fraudulent, but they could be legitimised by the issuance of a re-export permit by any country that would accept them. On account of this, several states, such as the United Arab Emirates, became substantial exporters of ivory. And, whenever one of the trading states was pressured into compliance, another 'hole' in the system was quick to develop.

The Management Quota System failed as a consequence of these clear inadequacies, resulting in a collapse of public confidence in the capacity for trade controls to work (Barbier *et al.*, 1990). These control system failures are not costless. It is essential that an effective control system is developed and implemented before all consumer confidence is permanently lost in the potentially constructive capacity of wildlife trade.

Another incentive system has been evolving in connection with, but not directly within CITES. This has occurred under the European Community Regulation 3626/82, which effectively requires (under Article 5) an import permit for all Appendix II species. More importantly for certain specified species (listed on Appendix C2 to the Regulation) there is an affirmative obligation on the exporting party to demonstrate that the export will not have a harmful effect on the population of the species in the country of origin.

This Regulation has acted to move the enforcement of sustainability from the producer- to the consumer-states (in regard to the EC), which is precisely what is required. This then allows for the imposition of the sort of 'conditionality' regime that was outlined in the first part of the chapter. The EC does in fact negotiate, on a country-by-country basis, the terms upon which it will remove that country's populations from Appendix C2. From this conditionality arises the possibility of state-by-state quotas created by the import permit obligation, and enforced by the EC.

As with the Management Quota System, this regulatory system provides for many of the necessary parts of a successful trade regulation mechanism. Its drawbacks lie almost wholly in its piecemeal approach. That is, optimal quotas require that the entire potential production of the species be considered. And successful incentives for sustainability require that all of the major consumers enforce conditionality; otherwise, supplemental conditions and enforcement by one consumer will often shift trade to others. Only one substantial reform is required: the EC regime should provide for development of quotas through joint consultation between producer and consumer states. In essence, however, the argument of this chapter is for the generalisation and extension of the EC regime across the consumer states (the US, Japan and the EC), with substantially increased involvement from the producer states.

The fundamental importance of all of these developments (ranching, quota and conditionality regimes) is that they represent a search for the 'middle ground' between Appendix I and Appendix II regimes. Although that search is only just beginning, its existence is demonstrative of the parties' awareness of the needs that the CITES system does not currently satisfy. CITES must continue to evolve to provide for an effective trade utilisation regime, rather than continue to act as the recorder of the process of species' evolution from Appendix II to Appendix I status. In order to do so effectively, there must be a clear target provided, otherwise the fiasco of the Management Quota System will only be repeated again and again. The control of wildlife trade is a potentially important instrument for the conservation of species, but CITES must develop a very different control mechanism if it is to contribute to the harnessing of these values. The concept of *quotas* that has developed within CITES, combined with the concept of *conditionality* that has been developed by the EC, is the preferred target.

8.7   SUMMARY – CURRENT AND POTENTIAL WILDLIFE TRADE MECHANISMS

8.7.1   The Current Control System

This chapter demonstrates two important points. First, there is a very definite and concrete economic meaning to be given to the concept of a constructive wildlife trade control mechanism. Second, the current CITES system does not, at present, conform to this meaning.

The clear and concrete meaning to be given to this concept is:

Optimal demand side policies   These consist of price-support policies which maximise the per individual value of a species, thereby enhancing the prospects of the species in the competition for scarce resources, and hence in the competition for survival.

Optimal supply side policies   These consist of trade-control policies which structure trade to make it conditional on demonstrated sustainable utilisation of the producer-state's natural habitat and the wildlife which it sustains.

CITES is not at present structured in the form of a constructive trade control mechanism. In general, an Appendix II listing does little or nothing to supply incentives to actually constrict the trade. In general, an Appendix I listing acts to withdraw demand from all suppliers of a species, the sustainable as well as the non-sustainable.

There is a fundamental need to understand the nature of the forces involved in a market, before there is any prospect for the development of constructive controls. CITES, despite the best of declared intentions, has failed structurally to provide these. Without constructive trade controls, there is an enhanced risk of species decline on account of both direct causes (overexploitation) and indirect (loss of habitat).

All of the proximate forces for extinction are closely linked to the quantities of efforts and resources put into the management and conservation of the species and its habitat. Overexploitation pressures must be met with well-financed management operations. Development pressures must be met with land-acquisition operations. The presence of both of these forces throughout the developing world requires the development of a trade control mechanism which *constructively* addresses both forces. One alone is insufficient.

The failure of existing controls, domestic and international, to create and to constructively channel rents from wildlife resources to their management and to habitat conservation is at the heart of both forms of extinction threats, direct and indirect. It is for this reason that CITES must evolve to address these needs.

8.7.2   The Direction for Change

CITES must evolve in the direction of a constructive trade control mechanism in order to address the joint threats of overexploitation and habitat losses as causes of species extinctions. In general, this will imply the need for the development of a regime of conditionality, where developed consumer states withhold purchases from developing supplier states unless:

Stock conditionality – the supplier makes and enforces a commitment to conserve specified habitat for the species concerned; *and*

Flow conditionality – the supplier's total annual product conforms to an externally enforced quota which is linked to the total sustainable offtake achievable from the designated habitat.

For many diverse resources, this form of regime would imply the creation of quotas in the range-states – only to the extent that 'Wildlife Management Areas' were designated as available for the species and then only in quantities which might flow from that habitat. If any of these states were not to designate areas as protected habitats, then they would not achieve 'certified supplier' status. When that status was conferred, it would only be for a quota that the designated area could sustainably generate. If activities incompatible with either 'designated habitat' status or the allotted quota occurred, then the certified supplier status would be withdrawn for a small number of years. This system would provide incentives for non-certified range-states to work toward sustainability, and for certified states to maintain it.

Then, it is the responsibility of the developed consumer states to enforce these quotas for the producers, creating and returning the rents from the resource to the managers of the resource. This would be done by the development of a rent-maximising joint quota, which is then divided between the certified producers in proportion to their 'sustainable quotas'. As with most wildlife resources, almost the entirety flow to the developed countries: the US, the EC and Japan. These states would have the responsibility of investing substantial amounts of resources in the monitoring of imports, in order to assure that only certificated diverse resources enter their borders.

Since it is very difficult to monitor for many diverse resources once they have been rendered into 'finished products', there are good reasons to simply disallow the re-export for manufactured goods made from these diverse resources. So long as manufactured goods are allowed to be exported, there is the possibility of disjunction between the manufacturer and the final consumer, and the former generally has less interest in the long-term survival of the resource.

The disallowance of reexporting may seem harsh, but it only implies that the manufacturer must move facilities towards the final markets. In this era of fairly fluid capital markets, it is not very unusual for manufacturers to do this.

Alternatively, there is also the option of direct 'green labelling' of products which are produced under a CITES certificate. Final consumer states would then match up labels with the quotas purchased by that manufacturer. This would require a good centralised data-handling centre, and very substantial commitments by consumer states to monitoring (of both retailers and imports). It is a hugely expensive alternative that cannot operate effectively in the absence of substantial commitments of funding, but it is the only option unless re-exporting is disallowed.

However, this alternative might also be preferred for another reason. The use of 'green labelling' at the individual consumer level would provide the consuming public with the important message that there are two kinds of wildlife products – those which are sustainably produced and those which are not. The former are pro-conservation and the latter are anti-conservation. At present, most consumers do not understand that such a separation is necessary; this has resulted in the withdrawal of substantial value from wild species, thus further threatening their existence.

The essence of a constructive trade control mechanism is to provide the means of making this distinction – between sustainable and unsustainable utilisation. The development of an option which recruited broad-based consumer support for such a mechanism would be a positive step. For this reason, the expensive step of multiple levels of screening might be pursued with regard to some wildlife resources.

8.8   COMPARING REGULATORY MECHANISMS

Chapters 7 and 8 have surveyed two very different regulatory mechanisms for conserving biodiversity. In essence, the two mechanisms surveyed thus far are based on two simple differences.

Direct vs indirect inducement   One mechanism focuses on direct transfer to induce the desired changes in the state's investment portfolio; the other focuses on investments in mechanisms which alter the decision-making framework of the state and thus indirectly impact upon its selection of portfolio assets.

Internal vs external regulation   One mechanism focuses on enhancing the benefits from diverse resources through direct subsidies to the state for internal management of diverse resource stocks; the other focuses on enhancing those benefits by investments in mechanisms which regulate the external markets for the products flowing from diverse resource stocks.

The choice between these two forms of intervention depends on the relative cost-effectiveness of the two. In general, this is a question of which form of monitoring (internal or external) is more cost-effective. This will depend in turn upon the number of potential producer-states that must be coordinated, and the costliness of monitoring imports rather than production. Often a *prima facie* case can be made for the cost-effectiveness of external regulation because those monitoring institutions already exist in the primary consumer-stsates (customs offices), and the number of natural producer-states of diverse biological materials is often very small (due to restricted ranges of many species).

In general, the instrument of direct inducement can never enhance state benefits from diverse resources by more than the marginal rate of return on capital (r). This is because a lump sum investment only assures a flow at that rate, and so this is the maximum perceived enhanced benefits that may be generated from that global investment, and hence the maximum rate of state investment that can be induced. The reason to consider the other instruments of inducement is the possibility of generating better rates of return through more resource-specific mechanisms.

There is one other possible reason to prefer a rent-appropriation mechanism to direct inducement transfers: the importance of institutional-isation. A rent-appropriation mechanism constructed around particular natural resources (such as an Ivory Exchange or a Medicinal Plant Clearinghouse) would require asset-specific investments by the Northern states, thereby creating some degree of assurance. It is difficult to create the same level of confidence in commitments to future flows of direct money transfers, because of the problem of time-inconsistency and the fluid nature of the assets involved. If a direct transfer mechanism is to have any effect upon Southern state asset portfolios, then it must be carefully constructed to remove all elements of future discretion from the Northern states with regard to the assets committed (by use of an independent trust fund, for example).

Finally, neither of these mechanisms address both halves of the problem of extinction. That is, the discussion to date has focused upon the induce-ment of Southern state asset selections; it has not touched upon the question of inducing optimal Northern state transfers. Neither of the mechanisms discussed above is able to resolve the problem of free-riding in the North. Rent-appropriation mechanisms may be able to contribute to the solution of this problem, *if consumers come to view purchases from the exchange mechanism as indirect investments in diverse resources stocks*. That is, as with 'free-range' products, consumers may be informed that consumption of a wildlife product will also pay for the natural habitat that the wildlife require, allowing consumers to express their willingness to pay for the latter by purchasing the former.

At present, however, this is the opposite of the information that con-sumers receive, and so it would be difficult to alter the signals so drastic-ally without substantial confusion in the marketplace. In the medium term it might be possible for rent appropriation to harness some of the stock-related flows to the Northern states, but it is important to address other instruments devised expressly for this purpose.

8.9   CONCLUSION

The solution to the global biodiversity problem is easily stated in theory. It requires only a system for the transfer of enhanced benefits from the global community to those owner-states investing in the supply of diverse resource stocks.

Chapter 7 demonstrated the complexity introduced into such a system when the dynamics of the exchange were considered. In this context, consistency and assurance became important characteristics of the institution-building process.

This chapter demonstrated the complexity involved in the implementation of a system devised to perform these functions. It focused upon a system which confers benefits through the creation of rents for the tangible goods flowing from diverse resource stocks, but the problems are general in nature. Any attempt to enhance benefits for investing states will fall prey to attempts by others to receive those benefits without incurring the costliness of the investments. Thus, a system for enhancing the benefits of investing states must be constructed in such a fashion as to effectively discriminate between investors and non-investors, and to target benefits only to the former. The catalogue of difficulties that this implies in the context of the wildlife trade indicates the difficulty of the task.

This is an important lesson to learn, precisely because the abstract simplicity of a problem of this nature is misleading. It is important to note that the true complexities involved in resolving many environmental problems are institutional in nature, and that it is the difficulty and costliness of their resolution that usually prevents or defers the solution to the environmental problem. The lesson to learn is that institutions must be created to evolve and change, with the increasing level of understanding and appreciation of the underlying problems to be resolved.

The twenty years of experience with CITES is a case in point. Even though the convention was initially based upon a clear misunderstanding of the nature of the fundamental causes of species-endangerment and extinction, it has slowly been evolving toward the creation of various forms of constructive trade regimes (especially quotas and trade certificate systems). It is important that this trend continues.

In short, the solution to the environmental problem of biological diversity losses is quite straightforward: it is the absence of a system of compensation for the non-appropriable services of diverse resources. However, underlying that problem and its solution is an institutional problem: the transfer of values for non-appropriable flows across boundaries in a multinational world.

It is this *international public good problem* that must be resolved. It is the decentralised (multinational) regulation of land-use conversions that generates excessive diversity losses, because the 'stock effects' are externalised. It is this same decentralisation that renders any solution costly to

implement. Transfers between the investing owner-states and the benefic-
iary consumer-states are difficult to arrange and enforce (as indicated in
Chapter 7), but the existence of other 'third-party states' attempting to
intervene in these agreements and intercept these transfers complicate the
transaction still further (as indicated in this chapter).

Therefore, the solution to this (as with any other) global environmental
problem is the evolution of a truly international institution in a multinational world. It is the creation and implementation of this institution that
is the most fundamental problem to be resolved.

9 International Regulation of Information: Intellectual Property Rights and Biodiversity Conservation

9.1 INTRODUCTION

The issue addressed in this chapter is whether there is a direct, market-driven response available for addressing the core problem of non-appropriable flows from diverse resources. That is, is there an instrument capable of rendering non-appropriable values appropriable? If there is, it would be of the nature of the regime known as *intellectual property rights* (*IPR*), and so the question becomes one of determining the applicability of this regime to the object of biodiversity conservation.

These two topics are necessarily related by reason of their roles in information-generation: biodiversity is a producer of information, and intellectual property rights regimes are the systems used to reward the producers of information. Biodiversity is a producer of valuable information because it is the product of the evolutionary process. Four-and-a-half billion years of evolution and co-evolution have produced a very particular pattern of biological life forms, and it is this pattern that encapsulates the history of that process. Any country that preserves in unaltered form the natural products of evolution is maintaining this history for the information that it is capable of producing. An intellectual property rights regime is the system of awards made to the producers of information in western societies, rectifying the otherwise 'public good' nature of information (Arrow, 1962). Therefore, it is only logical to consider the potential applicability of this system to the conservation of evolutionary product, i.e. biodiversity. Section 9.2 to 9.4 consider the nature of an intellectual property rights regime, and sections 9.5 and 9.6 its appplicability to conservation.

There are two subsidiary questions that are addressed within this chapter. The first concerns the possibility of introducing an inducement of host-state investments into diverse resources that is based upon traditional concepts of property rights. That is, is it possible to create an institution that renders a greater proportion of the flow of services from biodiversity appropriable

229

by the host state? A property right is of this nature; it renders a directed flow of benefits to the owner of the asset (Hart and Moore, 1990).

Irrespective of whether a property rights regime is feasible, the second important issue analysed here concerns the capacity to use the market to register consumers' preferences for the conservation of the currently non-appropriable components of the value of biodiversity. This is an important objective because it addresses the 'other side' of the diversity conservation problem, i.e. the problem of inducing optimal transfers from consumers to suppliers. The use of a market-based mechanism might supply some information regarding the appropriate scale of such transfers.

Therefore, the questions addressed in this chapter are threefold. First, the chapter investigates the nature of an intellectual property rights regime, and discusses its potential applicability to biodiversity investments. Secondly, it assesses whether a property rights regime is available to internalise the flow of services within supplier-state decision-making. Thirdly, the chapter investigates whether a market-based regime is available to register consumer preferences for some of the (informational) services of diversity. These last two issues may appear to be identical, but this is not the case; the argument of this chapter is that there are market-based approaches to internalisation other than 'pure' property rights. Interestingly, the analysis here indicates that the answer to questions one and three is affirmative, while the answer to question two is negative.

This result derives from the concept of a non-rival, but excludable, good (Romer, 1990a,b). A non-rival good is of the nature of a design or process, i.e. pure information. It is something for which its supply is undiminished in use, and thus its marginal costliness of consumption is zero. On the other hand, such a good may be made excludable, by virtue of government intervention, and it must be made excludable if there are to be incentives for investment in its production. A good example of a non-rival but excludable good is computer software; the marginal costliness of use is zero, but governments have put strenuous efforts into creating regimes for enforcing individual rights to the marketing of these products.

The form that excludability takes in regard to non-rival goods is unique. Since non-rival goods are without form or substance (as with information), it is not possible to demarcate boundaries around them. The dimension within which they exist ('information space') is not amenable to demarcation. It is instead necessary to substitute surrogate dimensions in which it is possible to mark boundaries, and use these to reward investments in information. In the case of software, for example, the rights would be given in the tangible manifestations of the underlying ideas, e.g. the written instructions, the design embossed on the computer chip or even the resultant display on the computer screen. All of these are surrogate, tangible dimensions being substituted for the intangible, underlying idea.

Some of the services of biological diversity are also non-rival goods, especially the information inherent within the natural diversity that is the product of the evolutionary process. However, in the case of natural capital-generated information (as contrasted with the human capital-generated varieties such as software), there has been no governmental intervention to effect excludability. These services of biodiversity remain non-excludable goods, and thus they suffer from predictable levels of underinvestment as a consequence.

This chapter demonstrates the manner in which non-rival goods (such as computer software) may be rendered excludable; this analysis constitutes a theory of the institutions known as 'intellectual property rights'. Secondly, the chapter demonstrates that the logic that supports government intervention for excludability applies equally as much to the analysis of natural capital-generated information as it does to human capital. There is no distinction, in logic, between the case to be made for the creation of intellectual property right regimes for biodiversity conservation than for any other informational investment (such as computer software).

9.2 FLOW APPROPRIATION MECHANISMS – 'INTELLECTUAL PROPERTY REGIMES'

The problem of appropriating international flows of wholly intangible services has been recognised and addressed for over one hundred years. The first truly international convention, the Paris Patent Union, was on precisely this subject; it attempted to create in 1868 an international mechanism for repatriating compensation to those who invested to generate information. Since that time a very substantial body of national and international law has developed around the idea of generating such flows, and these laws are known collectively as 'intellectual property rights' (IPR).

As will be demonstrated in this chapter, however, there is little in common between IPR and traditional property rights. The latter deal mainly with tangible commodities capable of exclusive possession and clear delineation. IPR deal almost exclusively with informational services, which are intangible and amorphous; they are not readily susceptible to either exclusive possession or delineation.

An IPR regime deals with this difference through the mechanism of 'surrogate rights', i.e. monopoly rights delineated in a dimension that is tangible in lieu of rights in a dimension that is not. The one acts as a surrogate for the other, in order to encourage investments in the intangible resource.

This chapter first describes the theory and application of surrogate rights (in the remainder of section 9.2–9.4) and then demonstrates the application of this theory in the conservation of biological diversity (in sections 9.5 and 9.6). Again, IPR regimes are important mechanisms to consider because they might potentially address the core of the problem of extinction, i.e. inappropriable service flows from diverse resource stocks, and/or they might provide a market-based mechanism for registering consumer preferences for biodiversity's services.

9.2.1   Failures in Property-Right Regimes – Unchannelled Benefits

A generalised statement of the purpose of an efficient decentralised management regimes is the encouragement of investments at the most efficient level of society, i.e. by those individuals who have the information and capability to invest most effectively in a particular asset (Hart and Moore, 1991). A very general statement of the nature of an efficient decentralised management regime is a mechanism that targets individuals making socially beneficial investments with awards approximately equal to the benefits generated.

This is the nature of a property rights regime, i.e. it is a mechanism for channelling benefits in a concentrated form through the hands of investors. It accomplishes this through the monopoly right known as a property right, which is a carefully delineated monopoly in the flow of goods and services from some specific asset. With the institutionalisation of the monopoly (and thus the assurance of the expectation that the state will invest to channel the asset's flow of benefits initially through the 'owner'), the individual is given the identical incentive framework as society's, i.e. the 'owner' will invest in the asset until the marginal benefit equates with the marginal cost.

For these reasons, property-right regimes are very useful mechanisms for inducing efficient levels of investments in various assets. However, some forms of assets are not amenable to the application of property-right institutions because their benefits are not readily channelled. In general, property-right institutions operate well if the flow of goods and services is densely concentrated in at least one 'dimension' at one point in that flow. Then it is possible to delineate and segregate the investor's flow from others', and hence to channel that flow. For example, most of the benefits from standard agricultural production (i.e. the produced commodities) are appropriable by the demarcation of exclusive rights in the land; the individual associated with a particular parcel of land (the 'owner') has the exclusive right then to the entire flow of benefits (produce) from that parcel. In this instance, there is a close correspondence between the total benefits generated and the benefits individually appropriated at one point in the process (i.e. at the point where the commodities are being produced on the land).

Many assets are not of this nature. Some have the characteristic that their benefits are instantaneously diffusive, so that investments in the asset generates benefits throughout a wide area. Others are of the character that their benefits diffuse rapidly and it is costly to segregate between beneficiaries and non-beneficiaries of the flow. This is the general nature of assets that are not easily subjected to property-rights institutions: the flow that they generate is too disorganised to be readily channelled. It is not easy to discern the point at which the award of a monopoly right will capture a significant part of the flow of benefits.

Consider again the first example of a globally-recognised public good, i.e. the information developed for industrial applications that was the subject of the Paris Patent Union. Investments in such information are not readily generated in the context of a property-rights regime, and so industry lobbied for novel forms of institutions (although confusing the issue by use of the same name).

Information is a global public good for two reasons. First, information as a product has an innate capacity for diffusion, as remarked upon in Arrow's Fundamental Paradox of Information. The paradox states that information is not marketable until revealed (because its value is unknowable prior to revelation), while the consumer's willingness to pay can be concealed after revelation of the information (because the transfer has already occurred). In addition, information is often revealed on the mere inspection of a tangible product within which it is embedded. Therefore, the mere act of marketing of a product created from useful information often releases that information to the world, rendering it far less valuable.

Secondly, it is extremely difficult to distinguish between information flows. This is because all information is built upon a common base (the common understanding that makes up all knowledge and language) and there are a multitude of potential pathways leading to the same conclusion. Therefore, an attempt to segregate between the path leading to one piece of information and the entirety of the remaining body of knowledge would imply an attempt to untangle all ideas back to the common starting point.

For example, consider the innovation of heat-resistant, resilient plastics sometime during the past fifty years. Most people site this innovation with the US space programme, where these substances were introduced in order to serve various purposes on the exteriors of spacecraft. However, on the use and observance of this information (i.e. the idea of durable uses for synthetic polymers), this idea diffused throughout the world economy. Soon, durable plastics appeared in the entire range of products, from automobiles to pots and pans. Even assuming that all of the consumer benefits from the use of these new products derived from the innovation at NASA, it would be very difficult to delineate clearly between the various uses (of a wide range of different polymers) or to trace their diffusion from the single point.

The nearly instantaneous diffusion of this idea throughout the economy demonstrates the difficulty of using property-right regimes to induce investments in a global public good, such as information. The benefits are never concentrated enough at one point in time to be channelled through the hands of an individual investor, because they diffuse so quickly and completely. For these reasons, other mechanisms than property right regimes must be used to encourage investments in assets that generate these types of flows.

9.2.2   The Role of Surrogate Rights Regime (and 'IPR' Regimes)

An alternative to a property-rights regime is a *surrogate rights regime*. Such a regime operates by channelling benefits to an investor from a monopoly right in a tangible good, as a reward for effective investments in an asset generating a non-tangible flow. In short, the surrogate right regime sidesteps the problem of non-appropriability by substituting a surrogate monopoly right (in a dimension that is suitably appropriable) for the impracticable property right in information. This is the nature of an IPR regime: it substitutes an appropriable flow for an inappropriable one in rewarding information-generating investments.

In order to understand how an IPR regime operates, consider again the example of the innovation of durable polymers. An IPR regime does not attempt to protect the investment of the agent who generated this fundamental idea; this would be impracticable for the reasons mentioned above. Instead, the IPR regime allows the agent to stake a claim in a carefully specified area of 'product space' where the idea is to be introduced. That is, the laws of patent do not offer rights in the idea itself ('durable synthetic polymers') nor to the entire range of products to which this abstract idea is subsequently applied ('all uses of resilient plastics from pots and pans to automobiles'). Instead, the applicant for a patent right must select a reasonable range of specific products that will make good use of the idea, and claim monopoly rights in the marketing of these. The inventor patenting the use of resilient synthetic polymers in pots and pans would not necessarily have any claim to a monopoly over their use in any other consumer goods, such as automobiles.

These benefit systems are not of the nature of property-right systems, but instead constitute a type of hybrid system for making awards to investors in information generation ('investors'). The general problem that the state must solve is how to create a cost-effective 'prize system' that will target efficient inventors accurately (in terms of identity and size of award) when the basic product is of such a nature that a pure property rights system is impracticable. It is not at all clear *a priori* that a surrogate rights regime is the most efficient institution for addressing this problem, but it is an interesting and internationally important example of a method for flow appropriation. However, there are several trade-offs involved in this particular solution to the problem, which will be explored in section 9.4, but initially the nature of a surrogate rights regime will be detailed.

9.3 THE INCENTIVES FOR INFORMATION PRODUCTION – A MODEL

Consider a Lancaster-type indirect utility space, where the ultimate objective of consumers is to maximise some measure of *indirect utility* (*V*) through the acquisition of various bundles of concrete products and product characteristics in *product space*. These products are instruments for the acquisition of utility (*U*), which is the set of individual goals and objects for that consumer. For example, the underlying goals might be 'transport between A and B' and 'demonstration of standing in community', while the product that might be first-preference as the instrument for attaining these goals might be 'new German sports car'. That is, each consumer has individual preferences distributed across product space, distributed in accordance with the capacity for a given set of products (*S*) to perform as an instrument to accomplish an individual's goals. The individual then maximises indirect utility (*V*) through acquisition of the optimal bundle of products (*V(S)*).

Consider also another distinct dimension, termed *information space*. Sets of information do not have a representation in product space, nor do they generate any utility (direct or indirect) by themselves. Rather, information is a third concept completely distinct from products or utility. In fact, information may be conceptualised as a 'bridge' between *U* and *V*, in that information defines the relationship between concrete goods and abstract goals. That is, with a given set of information (*I*), an agent is better able to understand how to obtain utility with the use of some or all of the available instruments (i.e. products and product characteristics). In essence, information acts as a shift parameter, making all levels of utility (derived from product consumption) relatively less costly to attain.

The Nature of Information

$$U = V(S;I) \quad \text{where} \quad V_{SI} > 0, \quad \text{but } V_I = 0. \tag{9.1}$$

equivalently $C_0 = C(V_0, p_S; I)$ where $C_{pI}, C_{SI} < 0$

where: $\quad C_0 \equiv$ the cost attaining V_0 given a specific price vector p_S
$\quad U \equiv$ utility (a function of individual goals and purposes)
$\quad V \equiv$ indirect utility (a function of product characteristics and information)
$\quad I \equiv$ information set for individual
$\quad S \equiv$ products and product characteristics

In this formulation, information is clearly a very valuable commodity. In its complete absence, there would be little understanding of how a given set of products or product characteristics could assist in meeting an individual's goals. It is knowledge, combined with tools (products), that makes the attainment of goals possible.

Consider now the incentives for the generation of information of this nature. A producer might, for example, combine information with various producer goods in order to satisfy consumers' fundamental preferences (for transport, nutrition etc.) at less cost. This information is often embedded in the producer's 'new' product, e.g. by its use of resilient synthetic polymers in the production of pots and pans. In fact, 'novelty' may be conceived of as the presence of information in combination with a set of concrete product characteristics.

A producer may decide to invest in information generation, but if it does so much of that information will be revealed on the marketing of a product using it (as with the resilient plastics), allowing competing producers to benefit equally from the information. In addition, other producers in entirely different industries will also receive the information and benefit from it, without compensating its producer. The widespread diffusion of the information on its initial marketing makes appropriation of its benefits impracticable; its value is too disorganised to engender individual incentives for its appropriation.

The insufficiency of individual incentives for investment in information is easily demonstrated. Consider a producer who is able to combine information with its products, making them less costly to produce (in relation to the functions that they perform). That is, in this instance, price is determined by reference to competitive products (i.e. competing methods for accomplishing the same purpose), but the information content of the good makes it less costly to perform the same purpose. An example would be an automobile manufacturer who decides to substitute resilient polymers in the construction of its bodywork; the consumer is able to accomplish its objects, but at less cost to the manufacturer. Will the producer have sufficient incentives to generate and use such information?

This producer is assumed to compete in a discrete product market termed product y (say, automobiles) as one of n producers, but information that is incorporated in y may be useful throughout product space (S) in the use of many different products. For example, the incorporation of these resilient plastics in automobiles may give the same idea to the manufacturers of pots and pans and spaceships. The producer must decide on the amount of information to combine with its product, given that information is costly to produce (at price p_I per unit) and given its individual appropriability.

Equation (9.2) provides both the socially optimal level of individual investment I_i^ω and the unregulated level of individual investment I_i^*. The unregulated level of investment is less than socially optimal for two reasons. First, with regard to the investing firm's own industry, it can expect to receive only the industry *average* benefit (y/n) for its investment. Since the information diffuses rapidly, all other firms in the industry achieve equivalent cost reductions and sales. The benefits of the investment are not channelled through this firm in particular, but averaged across the industry. Secondly, with regard to other markets in which the firm does not produce, there is no return whatsoever, even though the information diffuses beneficially across product boundaries. Therefore, decentralised decision-making regarding investment in information-generation will result in a socially inefficient level of investment on account of the diffusive nature of information.

Investments in Information Generation

$$I_i^* : \frac{\text{Max}}{I_i} \prod_i (y_i; I_i) \equiv \int_0^\infty e^{-rt} \left[p(y_i) y_i - c(I_i) y_i - \rho_I I_i \right] dt$$

$$\longrightarrow \quad I_i^* \quad : \quad \frac{c'(I_i)y}{n} + 0 \cdot \sum_{s \neq y \in S} c'(I_i) S = \rho_I$$

$$I_i^\Omega \quad : \quad \sum_{s \in S} c'(I_i) S = \rho_I \tag{9.2}$$

where : $\prod_i \equiv$ the present value of i's investments in

information (I_i)

$$y_i \equiv i\text{'s output of product in industry } y \ (n \text{ producers})$$

$$S \equiv \text{the set composed of all individual products } (s)$$

$$I^* \equiv \text{the decentralised optimal investment level}$$

$$I^\Omega \equiv \text{the globally optimal investment level}$$

There is one more important point to be made concerning the nature of information, and that concerns the nature of the generating process. It is usually assumed that the primary input into the information production process is human capital (together with a modicum of capital equipment). Then, the information generation process is a very simple one, involving the purchase of optimal quantities of suitably trained labour at its going price, and the price of information is equal to the price of this labour divided by its marginal productivity.

Information Generation (with human capital as input):

If $I = F(L)$,

Then $\rho_I = \dfrac{w}{F'}$

$$\longrightarrow I^*(L*): \ \underset{L}{\text{Max}} \int_0^\infty e^{-rt}\left[p(y)y - c(F(L))y - \frac{w}{F'}F(L) \right] dt \qquad (9.3)$$

Given this depiction of the nature of information, its production and value, it is straightforward to determine the optimal amount of information production that should occur. Information should be produced, and labour hired for its production, to the extent that the summation of the marginal benefits from further units of information produced exceed the marginal costs of their production. The remaining problems to be dealt with in the optimal production of information are two: first, the creation of incentive mechanisms to generate optimal investments in information, and second, the incorporation of information-producing inputs other than human capital.

9.4   THE THEORY OF SURROGATE PROPERTY RIGHTS

The problem of creating incentive mechanisms for the production of information is a very general one. The same problem has been analysed in regard to regulating the generation of information at different levels of a firm or distribution network. This analogue will be used in order to provide a private sector benchmark against which to compare the need for public sector institution-building (Matthewson and Winter, 1986).

9.4.1 The Nature of Surrogate Property Rights

Consider, for example, the problem of a manufacturer of a sophisticated consumer product (such as a personal computer) who wishes to market this product efficiently. The maximum number of sales will occur only if substantial amounts of information are included with the sale, e.g. informal demonstrations, lessons and instructions provided to prospective purchasers. For maximum effectiveness, this information must be provided on a decentralised basis (i.e. at the retail level), where the interface with the consumer is direct (in order to tailor the demonstration to the needs of that customer). However, these optimal investments will not occur on a decentralised basis on account of the inappropriability of retailer-generated information. That is, retailers who invest in the provision of these informational services (training of sales personnel, provision of demonstration rooms and equipment) will not be able to compete with those who do not make these investments, because consumers will have the incentive to acquire the (unpriced) information at one retailer and make their purchase at the other. Manufacturers need to construct mechanisms that channel the benefits from informational investments through the hands of their investing distributors, the identical decentralised investment-in–intangibles problem faced by the state in a more general context.

The private sector institution used to address this problem is the 'exclusive territory' regime incorporated within vertical distribution agreements. Such a regime, established by the manufacturer, provides that no other retailer shall be allowed to market the manufacturer's products within a carefully defined territory (from 50th to 195th street, say). This territorial monopoly right provides a local captive market from which to recoup the retailer's investments in informational services. Note that the desired investments are in a wholly inappropriable dimension – information – while the monopoly is allowed in an easily demarcated and segregated dimension – physical territory. The problems of inappropriability in the former dimension are addressed by allowing surrogate rights in the latter.

Analogously, the state needs to supply concrete rights in a dimension that can be demarcated, and product space serves this purpose. In the case of an IPR regime, a market is allocated by the specification of a concrete boundary in product space (as opposed to geographical space in vertical distribution agreements). An IPR regime acts to remedy this distortion in information-generation by granting monopolies in certain territories in product space. That is, in recognition of the impracticability of allowing monopolies in information, this regime instead allows monopolies in a range of products which incorporate this information. This supplies a remedy for the first problem of inappropriability (diffusion within industry) while supplying a premium to compensate the firm for the second problem of inappropriability (diffusion across industries).

An example of this is provided by the patent allowed to the innovator of the oversized tennis racquet. The actual innovation involved in that case was the idea that sports equipment sizes and shapes might be optimised; however, this concept (although widely implemented) is too abstract to be appropriable. Instead, the patent allotted to the innovator allowed exclusive marketing rights for all tennis racquets with head-size between 95 and 130 sq. cm in area. The tennis racquet actually marketed was of a single size, that fell in the middle of this territory; however the entire territory was allotted in order to create the monopoly rent.

Therefore, the idea of giving 'exclusive territories' as incentive systems for investment in certain assets may be applied even when there is a complete disjunction between the territory given and the asset requiring investment. The inducement of efficient investments requires institutionalised award mechanisms, and all institutions have their own forms of costliness. Surrogate property rights are clearly second-best types of solutions, but this is true of all institutions.

The idea that a 'property right' accomplishes a perfect match between asset and territory is illusory in every instance, giving rise to the prevalence of 'externalities'. To advocate 'well-defined property rights' is equivalent to advocating 'perfect competition'. It is important to recognise all property rights for nothing more than what they are: institutionalised incentive mechanisms for making awards (imperfectly) to investors. Surrogate property rights are substantively indistinguishable from all other property rights; they are both 'exclusive territories' operating as award mechanisms for beneficial investments. The difference is quantitative, in the quantity of externalities prevailing under the institution.

9.4.2   Surrogate Property Rights as Incentive Systems

Under an IPR regime, informational investments can be induced by awarding surrogate rights of a specific breadth and duration in the distinct (and concrete) dimension of product space (S). The breadth of a surrogate right is given by the set of products and product characteristics that are demarcated as being the investor's 'exclusive territory' (R_i). The length of a surrogate right is given by its established period of duration (T). The regulator's task, using this institution, is to determine the values of the variables breadth and length, in order to establish incentives for information-generation.

Equation (9.4) demonstrates the way in which the allocation of a clear and concrete territory in product space alters the decision problem of the agent contemplating an information-generating investment. In order to establish the 'prize' associated with an allocation of an exclusive territory in product space, assume that consumer preferences are distributed uniformly over that dimension (according to density function f). Hence, an allocation of a specific unit of territory in S (i.e. the exclusive right to produce products of those characteristics) to a given producer implicitly allocates a unit of consumer demand.

However, consumers whose preferences are sited at a particular point in product space are not captive at any price; the constraint in (9.4) states that the 'effective price' (product price plus the cost of distance from individual consumer's first-best product-τ) charged to the marginal consumer must equate with the 'effective price' of its nearest rival (located outside the allocated territory). That is, the consumer is willing to incur a premium of (τ) per unit of product space in order to have its first preference in product characteristics. Subject to this constraint, the producer can charge the monopoly price to its captive market.

The producer's incentives to invest in information are then enhanced by the availability of this captive market. The exclusive territory acts as a reward for such investments.

The Incentive System Established under a Surrogate Rights Regime:

$$I_i^{SR*} : \underset{I_i}{\text{Max}} \prod_i (p_i; I_i) \equiv \int_0^T e^{-rt} \left([p_i - c(I_i)] \int_0^R f(p_i + \tau s)ds - \rho_I I_i \right) dt$$

subject to : $p_i + \tau R_i = p_j + \tau(j - R_i)$ (9.4)

where : $\prod_i$ $\equiv$ the benefits channelled to agent i

      R_i $\equiv$ the territory in product space S allocated to i for period T

      j $\equiv$ the location in S of nearest competitor for i

      τ $\equiv$ the consumer cost of monopoly per unit distance

      T $\equiv$ the duration of the monopoly right

The creation of such surrogate rights generates incentives to invest in information, in both an *ex post* and an *ex ante* sense. It is an *ex post* award system, in that it creates a prize to reward demonstrably effective investments that have already occurred. The *ex post* nature of this award system operates as with any other public prize-awarding system. Contestants are judged and selected for the amount of 'public good' (external benefits) generated, and awarded prizes in compensation for those activities. The amount of the prize awarded, under an IPR regime, is equal to the difference in the expected returns from the enterprise with the IPR (i.e. with T>0 and R>0) and without (i.e. with $T > 0$ and/or $R > 0$).

An IPR system is also a fully *ex ante* system, in two senses. First, the prospect of such awards acts as an incentive system to generate informational-investments in anticipation of such an award. That is, to the extent that there exist expectations that exclusive territories will be awarded for all efficient investments, these *ex post* awards will be given full effect *ex ante*. However, it is not necessary to adjudicate awards for every successive investment. The IPR system also incorporates an explicit *ex ante* component in that it creates a territory within which future investments are automatically rewarded. That is, the award of the monopoly rent (irrespective of whence it originates) rewards past investments, and the award of the specific territory rewards effective investments occurring within that region in the future.

Equation (9.5) indicates the efficiency of the system after the award of an exclusive territory under a surrogate rights regime. As indicated below, these incentives cannot equate with the socially optimal unless either the information is sufficiently concentrated in the territory allocated, or the territory allocated is all of product space. Neither of these conditions can be met, because of the diffusiveness of information and the costliness of allocations. This is a purely second-best approach, based on the award of distortionary monopolies as prizes for effective investments.

Efficiency of Ex Ante *Incentives under Surrogate Right Regime:*

$$I_i^{SR*} : \int_0^T e^{-rt}\left(-c'_{l_i} \cdot \int_f^{R_i} f(p_i + ts)\, ds \right) dt = r\rho_I$$

$$I_i^{\Omega} : \int_0^{\infty} e^{-rt}\left(c'_{l_i} \cdot \int_0^S f(p_i + ts)\, ds \right) dt = r\rho_I$$

$$\longrightarrow I_i^{SR*} = I_i^{\Omega} \text{ iff : or} \quad (1)\ \ c'_{l_i} = 0 \quad \forall s \in R_i < s < S$$

$$(2)\ \ R_i \equiv S \qquad\qquad (9.5)$$

Despite these drawbacks, it is nevertheless apparent that a surrogate rights regime can be an effective mechanism for inducing informational investments. Given that any institution used for inducing investments in information will be costly, the important issue is the comparative costliness of the various institutions available for accomplishing this object. A regulator should invest in the institution of a surrogate rights regime to the extent that it generates returns (from induced investments) sufficient to cover its institutional costliness.

Equation (9.6) is a depiction of the regulator's problem in considering the institution of a surrogate rights regime. The establishment of an institution requires substantial initial investment, and it is thus represented as a 'stock variable'. The establishment of a surrogate rights regime thus involves the allocation of an adequate stock of management services to this institution. Equation (9.6) also demonstrates that the benefits from such a regime flow through both the information diffusing globally (π_D), and also through the benefits internalised to innovators via surrogate rights (π_S). The costs of the mechanism ($r\rho_M M_G$), including the distortionary costs, must be balanced against these gains.

Institution of Surrogate Rights Regime

$$M_G^{SR*} : \frac{\text{Max}}{M_G} \prod_G(I) \equiv \int_0^\infty e^{-rt}\left(\prod_S(R) + \prod_D(I) - r\rho_M M_G\right) dt \quad (9.6)$$

where: $\prod_G(I) \equiv$ the net global value of induced investments in information

$\prod_S(R) \equiv$ the value to investors in information of allocated terriories (R)

$\prod_D(I) \equiv$ the value to consumers of information generated

$M_G^{SR} \equiv$ global investment in institution of surrogate rights regime

Since information diffuses globally, equation (9.6) is a regulation problem for resolution by the global community. The analysis is identical to that involving consumer market regulation, discussed in section 7.5. Producer's investment incentives are being altered by means of the allocation of exclusive markets within consumer states. Market allocations have a much-reduced impact if other states produce and export competing goods. This explains why the Paris Patent Union was the first international convention concerning international resources, and also why allegations of piracy by non-member states have always been one of the major concerns of that convention. The more cooperation that exists between states in the institution of a surrogate rights regime, the more effective the regime.

Finally, it should be noted that the problem in eq. (9.6) is identical to the problem involved in the regulation of global biological diversity. Both concern the global community's regulation of a resource that generates an inappropriable flow of services. After a brief discussion of the efficiency characteristics of surrogate rights regimes in the next section, the application of this instrument to conservation goals will be discussed in section 9.5.

9.4.3   The Comparative Costliness of Surrogate Rights Regimes

A surrogate right regime is not an attempt to institute a first-best mechanism for information generation. A first-best approach would instead be of the nature of a periodic prize competition. This would consist of the award of *ex post* cash prizes to all effective innovators in the amount of the estimated global benefits generated. In this statement of the scheme, it satisfies the requirements of a first-best incentive system for generating efficient investments. In fact, such prize systems had been used as the primary inducement mechanism in the Soviet Union during the communist era, and the UK experimented with such a scheme in the post-war period.

The problem with the first-best ('inventors' prizes') approach is that it ignores the questions of institutional costliness. There are three categories of institutional costliness under which surrogate right regimes are likely to perform better than more general prize systems.

First, there is the costliness of valuing the benefits of informational investments. In a pure *ex post* system, this valuation must be accomplished entirely outside the market. With surrogate rights regimes, it is possible that some future investments will be rewarded through the monopoly over the captive territory. Therefore, there is some incorporation of the willingness-to-pay criterion involved in the surrogate right regime that is not incorporated in a pure prize system, and this may reduce the costliness of valuation.

Second, to a large extent, the surrogate right regime is merely a prize system with a lag between award dates, and the reason for the incorporation of such a lag concerns the importance of fixed costs in the process. If the process is undertaken at low cost and repeatedly, there will be a costliness involved in the inaccurate targeting of prizes (i.e. awards given to ineffective investors and the failure to give awards to effective investors). If the process is undertaken at high cost, some of these inaccuracies may be removed. However, the fixed-cost element of accuracy improvements suggests that there are gains to be had from less frequent awards (i.e. selecting some individuals for awards, and then allowing those selections to stand – with continued awards from the allocated territory – for a fixed period of time). It is the importance of accuracy that suggests that an infrequent, surrogate rights approach might be preferable over a frequent, prize-system approach. In essence, if a state incurs the costs of high accuracy, then it may be optimal to make the award an active and ongoing one (by the award of the exclusive territory).

It is rational to incur the costs of high accuracy because mistaken allocations are costly, especially in regard to the *ex post* element of the prize. This is because, whether the prize is in cash or territory, once it is awarded it cannot be transferred. Contrary to Coaseian analysis, the improper allocation of an intellectual property right cannot be remedied by subsequent transfer, precisely because it represents in part an award for effective *ex post* investments. If misallocated but subsequently transferred, the territory may be properly allocated but the monopoly rents (based on past uses) stay with the initial holder.

Therefore, the primary reason to consider the use of a surrogate right regime is this fixed cost element to the selection mechanism. The greater the costs of accurate selections, including the costliness of inaccurate selections, the more an occasional, dynamic mechanism will be preferred.

Finally, a surrogate rights regime involves the creation of a substantial amount of international infrastructure, in the form of a legal process for the evaluation, allocation and demarcation of market territories. This is an important step toward the institutionalisation of the award system, and it creates the assurance required to generate investment. It is probably for this reason alone that these regimes are known as 'intellectual *property*'.

9.5 THE APPLICATION OF INTELLECTUAL PROPERTY RIGHTS TO BIODIVERSITY CONSERVATION

The entirety of the theory developed in this chapter applies directly to the problem of biodiversity conservation. This is because investments in stocks of diverse resources (species and habitats) generate not only tangible goods and services, but also intangible ones (specifically, insurance and information). On account of the diffusiveness and non-segregability of these services, it is not possible (under existing institutions) to channel these global benefits initially through the hands of individuals living within their host states.

However, this flow of information is only maintained by way of investments in diverse stocks by individuals in the host states. If these diverse assets are not included within state portfolios, then the flow of these services will cease. It is as important to reward investments in natural capital that generate informational services as it is to reward investments in human capital-generated information. The base problems are identical, only the physical character of the asset involved is changed.

9.5.1   Biodiversity as Evolutionary Information

In the context of the model developed in section 9.3, the only element to add is the potential production of valuable information from inputs other than human capital. As has been indicated at several points in this book, this is the essential value of biological diversity – its informational content. It must be recognised that human capital alone may not be capable of producing all important and valuable information. There is also a base biological dimension which generates information $I(R)$.

This biological dimension is the evolutionary process which, through biological interaction and the process of selection, generates communities of life forms that contain substantial amounts of accumulated information. Because the competition for niches is constant and pervasive (occurring at all levels), the naturally-evolved life forms contain biological materials which act upon many of the species with which they share the community. A community that has co-evolved over millions of years contains an encapsulated history of information that is not capable of synthesisation.

Supplanting a naturally-evolved habitat, and slate of species, with a human-chosen slate may confer tangible productivity gains, but it also removes the information that was available from that community. The information from co-evolution, the product of the evolutionary process, is lost with the conversion.

Therefore, the conversion of the last remaining unconverted natural habitats equates with the retention of this evolutionary product, i.e. the information generated by co-evolution. The mere existence of this habitat represents information production, in the sense of the retention of an otherwise irreplaceable asset. Valuable information may be produced by invesments in natural capital as well as through investments in human capital.

9.5.2   Incentives for the Conservation of Biodiversity

Consider how the global community can use a surrogate rights regime in order to conserve optimal biological diversity. The global community is faced with the same regulatory problem as listed above in eq. (9.6), but with slightly different dimensions involved. In eq. (9.7), the benefits to supplier-states (Π_S) flow primarily from allocations of exclusive markets in consumer-states (R^P), while the benefits to consumer-states (Π_D) flow primarily from the retention of natural habitats in supplier-states (R^N). This is analogous to the object outlined in eq. (9.6); the problem is to invest in institutions to maximise the aggregate benefits from informational production for both consumers and suppliers.

International Regulation of Global Biological Diversity

$$M_G^{SR*} : \int_0^\infty e^{-rt} \left(\sum_S \prod_S (R_S^P) + \prod_D (I(R^N)) - r\rho_M M_G^{SR} \right) dt \qquad (9.7)$$

where: $\prod_S$ $\equiv$ the value to supply states of allocated terroitory in product space (R^P)

$\prod_D$ $\equiv$ the value consumer states of information flowing from natural habitat (R^N)

M_G^{SR} $\equiv$ the global community's investment in surrogate right regime

In essence, the global community is allocating territories in Northern product markets (R^P) in exchange for the conservation of designated territories (R^N) in Southern natural habitats. These rights constitute both *ex post* awards for past effective investments, but also *ex ante* awards to encourage investigations for further useful information in those territories. Such awards function in precisely the same way as intellectual property rights in encouraging investment, except that in this case the regime is focusing upon natural-resource (rather than human-resource) generated information.

As with the mechanisms discussed in Chapter 8, surrogate right regimes apply international management of product markets to the inducement of investments in domestic resource commons. The state incentive structure, under a surrogate right regime, appears in eq. (9.8). The owner-state of a significant area of unconverted habitat receives enhanced benefits (flowing from the global regime) from all investments in the management and preservation of that habitat.

Investment Incentives for Supply State under Surrogate Rights Regime

$$x_G^{SR*}, R_S^{SR*}, M_S^{SR*}: \underset{x_S, R_S^N, M_S}{\text{Max}} \prod_S(p;\cdot) \equiv \int_0^T e^{-rt}\left((p - c(x_S, R_S^N, M_S)) \left[\int_0^{R_S^P} f(p + \tau s)ds\right] \right) dt$$

$$\longrightarrow \frac{\partial \prod_S}{\partial(\cdot)} = \int_t^T e^{-rt}\left[(-c_i') \int_0^{R_S^P} f(p + \tau s)ds \right] dt \text{ for } \partial x_S, \partial M_S, \partial R_S^N \,. \tag{9.8}$$

The supplier states now have incentives to invest in their diverse resources. They have two sets of incentives. States have the incentive to maintain their resources and investigate them in order to be awarded product-market territories, and states have the incentive to continue to invest in their diverse resources in order to generate new information useful in respect to their market allocations.

The extent of these incentives depends entirely upon the breadth and length of the awards. The breadth is determined by the extent of product-market allocation, and the length is determined by the duration of the allocation. Different criteria should be used in determining breadth and length. Length should be determined primarily with regard to the costliness of selection and allocation, as discussed in section 9.4. Breadth should then be used to establish the desired amount of the award.

As with any surrogate right, it is a more efficient instrument if it is able to capture a large proportion of the information's value in the product space allocated. However, by the definition of information (i.e. its diffusive nature), this is not generally possible. It must be recognised that the ultimate object of any surrogate right (intellectual property right) regime is to accurately target prizes to efficient investors in information generation, and that it is only institutional costliness that warrants the use of surrogate dimensions for this purpose. Therefore, intellectual property rights can be an effective instrument for the generation of a flow of value to states investing in the conservation of biological diversity. This instrument should be considered with the others as a potentially cost-efficient method for conservation.

9.6 INTELLECTUAL PROPERTY RIGHTS IN BIODIVERSITY CONSERVATION – INFORMATIONAL RESOURCE RIGHTS

A surrogate rights regime based on natural resource investments would allocate product-market territories in response to effective natural habitat investments. Effectiveness would be demonstrated initially through the establishment of an effective conservation and investigation programme, shown to be capable of identifying and conserving useful life forms. When such a programme is established, the potentially useful life forms need to be tendered to an international natural resource panel, together with a suggested range of uses for the species. The panel would then determine several issues: (1) whether an award should be made to the programme; (2) what the product characteristics are that might be generated from the natural resource; (3) what breadth of award in regard to these specific product characteristics are to be lodged with the programme.

The determination of issue (1) depends upon the perceived usefulness of the natural habitat and programme applying for the award. Issue (2) turns on the derived chemical usefulness of the specific products tendered. Issue (3) will consist of a determination of the specific royalty payable to this programme by companies operating in the zone of the protected product characteristics. Of course, the length of the award is a separate, institutional issue determined uniformly by the costliness of the selection process. In this fashion, a constant source of funding for natural conservation could be maintained in a fashion that both links funding to usefulness and also creates incentives to invest in the source habitat.

9.7 CONCLUSION

The answers to the issues outlined in the introduction are now apparent. It is clear that there is no distinction in substance between investments in information-generating diversity and other information-generating assets (such as other research and development activities). Therefore, there is no logical reason why intellectual property right regimes should not be applied to the conservation of biological diversity.

The major advantage that IPR regimes demonstrate is their capacity to bring market-based preferences into the calculations concerning the value of diverse resources. Although the link is tenuous and institution-dependent, these arguments apply equally to all 'intellectual property rights'. For example, it can be no more difficult to value and assign rights in the services rendered by natural diversity than it is in regard to the 'look and feel' of a computer–user interface (patent granted to Apple Computers). The analogy is direct between the computer software industry and biodiversity conservation. Both 'industries' produce information – one in the variety of the code and one in the variety of the life forms. Both forms of diversity are useful – one in the operation of a computer and one in the operation of the biological production system. But the values from both forms of informational services are largely inappropriable (after first sale) unless governments make a concerted attempt to reward the producer.

The primary difference between the application of IPR regimes to software versus biodiversity is the identity of the rewarded producer; a biodiversity-related regime would produce largely North-to-South flows while the existing regime produces substantially North-to-North flows (and substantial South-to-North flows). Possibly for this reason, there are massive resources being spent on the reform of the international IPR laws concerning software protection, but little interest in investments in the creation of IPR in natural resources.

However, if this is the case, then this myopic view of Northern self-interest concerning international IPR regimes completely misses the point. The rationale for an international institution should be the appropriation of the 'gains from cooperation', and in the case of biodiversity, the gains from cooperation inhere when Northern states transfer funds to the South in return for the Southern states' conservation of diverse resource stocks. If properly calibrated, these transfers will be made in a fashion that will reward and induce compensating investments in diversity. For this reason, IPR regimes for natural resources should generate a net gain for Northern states, although the flow of funds under their auspices will be unidirectional North-to-South.

The solution to the global biodiversity problem requires the construction of mechanisms to address 'both sides' of the problem: optimal supplier state conservation and optimal Northern state payments. It cannot be assumed that whatever level of funding that happens to emanate from the North is optimal in any sense. It is therefore equally important to invest available funds in the development of incentive systems that induce optimal behaviour in the North, as it is to invest in the development of such systems for the South. It is for this reason the instrument of IPR regimes for biodiversity conservation is especially appropriate.

However, all three of the instruments discussed in the preceding chapters have much in common. They are three different methods for attaching an internationally-created 'global premium' to the local benefits received from diverse resource management. The 'direct' approach of Chapter 7 involves the payment of this premium for the conservation of diverse resource stocks at low-intensity uses. The 'trade' approach of Chapter 8 attaches this premium to the tangible, appropriable components of the flow from these diverse resource stocks. The 'IPR' approach of this chapter hives off a percentage from the commerce in derived tangible products for compensation of the intangible inputs into its production that flow from diverse resource stocks.

Once again, the real difference between the three different approaches is an institutional one. Each constitutes a potential conduit for transferring the values placed on low-intensity uses of diverse resources back to the owner-states, i.e. each is a candidate for the creation of *internationally transferable development rights*. The direct approach accomplishes this through periodic payments in return for these rights. The regulated trade approach does this through altering the rates of return to low-intensity uses of diverse resources (as opposed to conversion). The IPR approach does this by creating a hybrid system for making awards to compensate both the past and the future retention of diverse resources. The essential solution to the problem is the same (i.e. the attachment of the 'global premium'); only the institution used to perform the task is distinct in each of these cases.

Therefore, it is important to recognise that, irrespective of the instrument chosen, the essential task remains the same. It is necessary to create an international institution or institutions capable of measuring, allocating and enforcing the global premium required to conserve biological diversity.

10 The International Regulation of Extinction – Conclusions

10.1 INTRODUCTION

The object of this book has been the development of a framework for the explanation of the nature of the biodiversity problem, in such a manner that the explanation might itself suggest its own solution. To this end, it is necessary here to recount only that the problem of biodiversity is a clear example of the divergence between the locally and the globally optimal. Due to various externalities within the process, each state has the incentive to convert its diverse resources to a slate of sameness. For this reason, over the past ten thousand years, the amount of global diversity has been slowly converging upon a small slate of specialised species.

As the final refugia of the world's diversity are reached, the rate of extinction will escalate geometrically. Since diversity is the essential component of evolutionary product, the potential costliness of its loss is unlimited. The global process of convergence must be halted well before this point is reached. This is the problem that we must resolve.

If it is accepted that all human societies have equal rights to development, then there is a unique solution to the suggested problem. This is the creation of international institutions capable of channelling the values of diverse resources to their host states. The creation of such alternative pathways to development maintains both the important natural asset as well as the essential human right. It is the fostering of development compatible with diversity, by means of investments by the already-developed states.

The conclusions within this chapter describe the nature of the regimes that might be developed to implement this solution. Section 10.3 constitutes a guide for the development of protocols to the international convention on biodiversity, that would be consistent with the solution of the problem outlined in this volume. Equally important is the understanding of the impacts of existing policies on the outlined problem. These impacts are outlined in section 10.2.

Clearly, there are many other problems that might be included on a wide-ranging rubric such as 'biological diversity'. The intent here is not to attempt solutions to all problems relating to biological resource management, but rather to attempt the solution of one very specific problem with very important implications for this resource system: the management of the global conversion process. This is the problem of biodiversity losses discussed here, and the concepts outlined here are directed to this one specific purpose.

10.2 THE INTERNATIONAL REGULATION OF EXTINCTION: EXISTING POLICIES

The international regulation of extinction must be addressed to the fundamental cause of diversity decline. This means that biodiversity policies must be directed to influencing decisions by owner-states concerning conversions. These states must be convinced that development is possible in the context of retained diverse resources, otherwise the conversion of these resources will occur. Therefore, the policies suggested by this approach must be directed to the channelling of the values of diversity through the owner-states.

It is also important to consider the effect of existing endangered species policies that are not focused on the fundamental causes of extinction. Some of these policies are developed to instead address the consequences of conversion decisions, e.g. overexploitation. Policies that have been devised to address overexploitation often have little or no positive impact on the forces causing extinctions. The existing extinction policies must be overhauled in order to redirect their focus to the more fundamental causes of extinctions.

This section surveys existing policies, and how they impact upon the conversion process that effects extinctions. The first two sections concern prevailing policies concerning information appropriation and agricultural insurance, and how they relate to diversity-generated services. The last section discusses the existing international wildlife trade system, especially as it performs as a policy for rent appropriation.

10.2.1 Information and Intellectual Property Rights

The purpose of the legal regime known as intellectual property rights is the protection of information-generating investments. As discussed above, there are no market-based incentives for investment in assets that generate valuable information because of the inability to appropriate the benefits that result from that effort.

Intellectual property rights address this problem by creating a link between investments in information creation and flows of revenue from the use of that information. Specifically, the government awards a mono-poly right for the marketing of a particular range of tangible goods in order to compensate a person for investments which ar obviously effective in the generation of information. These protected markets (or intellectual property rights) then provide the basis for compensating both the past informational investments, as well as any future ones made in that area.

Intellectual property rights are used to reward informational investments in a wide variety of markets. For example, in the area of computer software, entrepreneurs invest efforts in the creation of effective software packages which they would then like to sell to the public. These software packages are essentially lists of 'instructions' that cause a computer to perform a valued function. That is, they represent almost pure information. For this reason there is no market-based mechanism that is capable of channelling the value of these recipes to their discoverers, no matter how highly valued the information might be. Once the information is marketed to a single person, it is potentially available to all. In effect, the initial marketing of the 'recipe' releases the idea, which is then free for use by all (through pirating or even the mere communication of the idea and imitation in a slightly different form).

Recently, legislature and courts have been struggling to create more en-forceable rights in software, because of the perceived value of these invest-ments. (For example, the Copyright, Designs and Patents Act of 1988 (UK) and the Semiconductor Chip Protection Act 1984 (USA).) They have been extending rights back toward the 'idea' itself rather than the particular form it takes. This is accomplished, for example, by giving exclusive rights in the marketing of anything with the same general 'look and feel' of the protected creation, rather than its specific manifestation. (*Apple Computers* v. *Franklin*, 714 F.2d 1240 (1984)). The law is not yet settled, but the direction of the efforts is clear; the developed countries are putting tremendous efforts into creating protection for this industry because of a belief in the value of the information that it can generate, if its markets are protected.

Analogously, investments in the retention of lands with diverse resources and the exploration of their usefulness also generate information. The diversity that exists in the wilds of the world contains the information necessary for insuring our crops, health and lifestyles. The discovery of new uses for these natural resources is equivalent to the identification of a new recipe for a useful function. A good example is the discovery in 1963 that chemical combinations found in the Rosy Periwinkle, a medicinal plant from Madagascar, are useful in the treatment of several forms of cancer.

It would appear to be of little consequence for the purpose of policy-making whether the valuable information were to be derived from an understanding of plants or computers. Explorations regarding biological materials can be as useful as explorations regarding machinery. For example, the health-related benefits from the chemicals identified from the discovery of the Rosy Periwinkle generate an estimated $88 million per annum. However, almost without exception, current governmental policies expressly disallow protection of discoveries concerning the usefulness of naturally-occurring resources.

It is not possible to claim rights in the information generated by the identification of new uses for naturally-occurring substances, irrespective of the value of the discovery. For example, Article 53(b) of the European Patent Convention states that no protection is available for 'plant or animal varieties or essentially biological processes for the production of plants or animals'. Similarly, in the landmark US decision of *Diamond* v. *Chakrabarty*, the Supreme Court declared that it was indeed possible to claim patent rights in live plants and animals, but the basis for awarding a patent in a living organism was that: 'the patentee has produced a new bacterium with markedly different characteristics than any found in nature' ... *His discovery is not nature's handiwork, but his own; accordingly, it is patentable subject matter under patent law*' (emphasis added).

There is no conceptual basis for the distinction that is being drawn between informational investments in natural versus manmade resources. There is also no practical basis for distinguishing between the two. The creation and protection of a system of intellectual property rights in discovered usefulness of naturally-occurring resources would certainly be no more difficult than the attempts to do the same with computer software. The basic resource – information – is identical, and the natural resource is probably more easily defined and contractually licensed than is the software product. In essence, once the right is recognised, the impacts would be twofold: first, there would be an exclusive right to market discoveries derived from analysing natural resource usefulness; and secondly, there would be a subsidiary right to license 'prospecting' for useful biological materials in existing natural habitats.

Not only has the international community refused to recognise the first right above, it has also taken steps to reduce any residual value that might remain for prospecting in natural habitats. This is most evident in the policies of the international system that has been created for the purpose of preserving plant genetic diversity, the International Board for Plant Genetic Research. The IBPGR strategy for conserving genetic resources has been to remove plant germplasm from its natural site to a germplasm bank, and then to provide 'free access' to the collection (Juma, 1989). This policy has effectively disenfranchised the host countries from their own biological assets, and greatly reduced the value of prospecting in the country itself. The value of the exploration rights that have been denied is indicated by the number of requests for access to these gene bank collections.

10.2.2   Insurance and Crop Insurance Programmes

The provision of insurance is an important aspect of social welfare. As surveyed above, biological diversity serves to provide insurance with respect to crop productivity. The failure to return this value to the countries generating it results in a lack of incentives to maintain the service.

It is important to recognise that there are substitutes for certain insurance services of biodiversity. Insurance can be provided by 'pooling' risks across a large enough group. For example, food producers can self-insure through the market by purchasing crop insurance against pests, weather and other hazards. This market-provided insurance is equally effective in reducing certain individual risks through aggregation. By means of pooling, each individual farmer takes a small loss with certainty (the insurance 'premium') rather than the chance of a large loss. This reduces the risk experienced by each individual farmer, which is one of the roles of a system of insurance.

However, there are certain risks that cannot be insured through the market. This is the case when risks across the insured group are correlated; i.e. if, when they occur for one of the insured, they are more likely to occur for the others. These are sometimes known as 'environmental' risks because of their general nature. Pooling does not operate to reduce these risks because all of the participants in the pool are likely to be afflicted simultaneously. This type of insurance cannot be provided through the market.

The risks to crops are precisely of this nature. As the environment changes (climatically, biologically or otherwise), there is the chance that the then-utilised specialised strains will not be well-adapted to the new environment. Change within the environment is its inherent nature, and some of these changes occur on vast scales: regionally, nationally and globally. In general, it is not possible to insure against these risks through the mechanism of pooling.

Nevertheless, some governments are attempting to substitute market-provided crop insurance for that which can be provided through diversity. For example, the US Federal Crop Insurance Corporation currently subsidises the purchase of market-based crop insurance by 30 per cent. During the 1980s, the US government spent $3.7 billion on attempts to make this market work (Table 10.1).

Table 10.1   US government expenditures and subsidies on market-based crop insurance policies (000s of US dollars)

Year	Subsidy	Losses	Admin.	Reinsur.	Total
1980	46 995	30 471	87 000	3 667	169 417
1982	91 990	132 250	115 000	23 138	363 546
1983	63 669	297 971	95 000	35 603	492 946
1984	98 296	205 314	98 000	78 887	480 364
1985	100 224	242 438	97 000	102 888	542 270
1986	88 043	233 806	95 000	97 711	515 352
1987	87 536	4 669	72 000	97 418	262 099
1988	107 830	585 678	76 000	121 545	891 830
Totals	684 538	1 731 597	740 000	560 583	*3 717 824*

Source: 'Report of the Commission for the Improvement of the Federal Crop Insurance Programme', FCIC: Washington, DC (1990).

By means of this subsidy, the US government is encouraging the substitution of a pooling-based system for a diversity-based system of insurance on a nationwide basis. However, since the risks to crops are necessarily correlated, this system cannot operate through the market alone. It will always require a governmental subsidy. This subsidy then acts to reduce the demand for diversity. The end result is that the system for determining the optimal amount of biodiversity is again biased downwards.

The most important problem is one of so-called 'moral hazard', i.e. the problem that insured people do not take adequate precautions against possible hazards precisely because they are insured. A crop insurance programme can lead to complacency abut potential large-scale environmental risks, because the small-scale risks that occur more frequently are adequately dealt with. However, when an environmental risk of national or global impact occurs, there is nothing that a pooling-based programme can do about this. Biodiversity-based insurance services are being substituted for, even though there are many events for which there are no substitutes for biodiversity.

In short, governmental policies should encourage payments for the already-provided biodiversity-related crop insurance. Governmental policies should not subsidise the substitution of market-based crop insurance, as this is necessarily ineffective and self-defeating. Insurance that can only be provided through the conservation of diversity must be obtained through a policy directed to this end.

10.2.3   Diverse Resources and Wildlife Trade Regimes (CITES)

This source of policy failure is far more obvious and straightforward than the previous two. The decision to convert natural habitats will depend upon the landowner's relative valuation of the resources in the two states: wildland or agricultural land. Most governmental policies on the subject operate to reduce the value of wildlife while subsidising the value of specialised agricultural commodities.

There has been much written on the subject of subsidies to agriculture in the OECD countries. In the developing world, the problem is as severe as it is in the developed. For example, it has been shown that in the case of the Brazilian Amazon, it would often be financially infeasible to convert the forest to cattle ranching in the absence of prevailing governmental subsidies (Browder, 1988).

Very likely, the less well-known and more important problem is the unwillingness of governments to place any significant effort into capturing the substantial values of already-produced wildlife resources. For example, it has been demonstrated that several Asian states capture only a small portion of potential rents (10–30 per cent) from the marketing of tropical timber (Repetto, 1988). The same is true with regard to most animal species. Most of the African states captured only a small proportion of the value of exported elephant ivory (5–15 per cent) (Barbier, Burgess, Swanson and Pearce, 1990). Parrots from Irian Jaya are exported for $2 and then sold in the US at $500 wholesale (Swanson, 1991b). This is a very general phenomenon in the world wildlife trade.

It is self-evident that these governmental policies are based in an underlying belief that development is not possible within the context of diverse resources. These governments are indicating that they believe that conversion must occur as a prior condition to development of the territory.

The international regime currently in place to regulate the trade in diverse resources should be built upon the premise that this perception must be reversed; development must be seen to be compatible with diversity. The existing regime was not built upon this foundation; it was developed to address the consequences of decisions not to invest in diverse resources rather than the causes of these decisions.

The Convention on Trade in Endangered Species (CITES) was adopted in 1972 and provided a single international instrument for the regulation of trade in diverse resources: the trade ban. In effect, CITES provides that a species listed as 'endangered' (as an Appendix I species) should be banned from international trade. All other trade proceeds virtually unregulated.

This sort of policy accords well with a theory of extinction based upon 'overexploitation'. Overexploitation is viewed, in this model, as deriving from the application of market pressures to wildlife (i.e. uncontrolled) species. When the market pressures threaten to overwhelm the species, the policy reforms suggested by this model are to criminalise the supply of and demand for the resource. In this way market pressures are relieved and the resource is allowed to restore itself.

Of course, this approach to extinction policy does nothing to address the underlying causes endangering the species, which is the unwillingness of the owner-state to invest in the management of the pressures on the resource. In fact, actions rendering the resource less valuable reinforce the perception of the owner-state that the resource is not worthy of invest-ment. The initial state (of overexploitation) was the result of this percep-tion, and the 'endangered species' laws developed under CITES have only acted to bolster this viewpoint.

Therefore, the existing legislation regarding species decline does not address the fundamental causes of endangerment, but instead deals only with the consequential. So long as these subsidies to specialised agri-culture (in the developing countries) and bans on wildlife commodities (in the developed countries) are in place, the perceived relative value of diverse resources will be reduced. An international regulatory policy for conserving biodiversity must reverse or reform these policies.

10.3 THE INTERNATIONAL REGULATION OF EXTINCTION: THE NECESSARY REFORMS

The international regulation of extinction must be directed toward encouraging those undeveloped countries with substantial quantities of diverse resources to consider alternative pathways to development that do not require the conversion of their resources. This implies the develop-ment of policies for enhancing the benefits that these states receive when they invest in their diverse resources (as opposed to disinvesting in these resources).

There are three methods available to the international community for conferring enhanced benefits upon investing states. First, the international community might simply confer direct subsidies upon those states retain-ing stocks of diverse resources intact through low-intensity utilisation; the form of the international agreement that would divide the use of a diverse habitat between the international and the local communities is known as an international franchise agreement. Second, the international community might intervene by means of regulating the trade in the tangible commodities that flow from diverse habitats in order to maximise appropriable rents; this would occur through reforms to the international wildlife trade regime. Third, the international community might intervene by methods directed to reach the core of the problem, i.e. by the creation of mechanisms for the channelling of the inappropriable values of diversity (insurance and information) to the states that supply them. This would imply the development of regimes analogous to intellectual property rights for purposes of diversity conservation. This section summaries the nature of all three of these institutions.

10.3.1   International Franchise Agreements

If the international community wishes to acquire the 'development rights' of individual states with regard to particular diverse resources, then it will be necessary to do so within the context of a dynamic international agreement. That is, there will have to be some overarching framework that provides for the systematic compensation of the owner-state at the end of each period that the diverse resources remain unconverted. It is only through the creation of such a dynamically consistent scheme of payments that the agreement concerning development rights can be made enforceable.

An international franchise agreement is the contract form that will allow for this system of governance. A franchise agreement is a contract that provides for a limited term of use by the holder of the franchise, subject to restrictions placed upon that use for the benefit of a third party. In this context the franchise agreement would divide the rights of use of a territory between the international community and the local community, with the owner-state standing as the intermediary within the contract (i.e. the owner-state is a party to the contract with the local commuity and with the global community). Both sides of this three-party agreement are made enforceable by means of the periodic payments made by the global and local communities in exchange for their partition of the rights of use; that is, the owner-state has an incentive to enforce the agreed partition of rights in order to receive payments from both sets of users.

The international franchise agreement option is flexible enough to allow for virtually any manner of division of rights of use between local and global communities. If the global community is willing to outbid the local community for all rights of use, then the area may be effectively 'zoned' as wilderness territory. If the global community bids are relatively small amount (compared with the area's use value), then the area may be zoned for all uses other than clearing and conversion. As mentioned above, whatever the partition between global and local communities, it is automatically enforceable because the partition is based upon an auction of the rights to use the land. The state must effectively enforce the agreed partition in order to receive payments from both sides to the franchise agreement.

This first route to the international regulation of extinction is based entirely on the idea that the flow of services from diverse resource stocks is directly related to the quantity of stocks remaining. Then, the optimal provision for diversity's services will equate with the conservation of diverse resource stocks. If this is the case, then the form of agreement by which the international community may acquire 'development rights' within owner-states is the creation of international franchise agreements.

10.3.2 Wildlife Trade Regimes

An alternative route for providing enhanced benefits to those investing in diverse resource stocks is to provide an enhanced return to the appropriable goods and services that flow from these stocks. These returns would then be attributed, within the decision-making framework of the owner-state, to the investments in diverse resource stocks, and this would create the incentives for their retention.

The development of a 'premium' attaching to diverse resources would be straightforward. This could be implemented through exclusive purchasing agreements between consumer states and producer-state cooperatives. The members of the producer cooperative would then restrict joint production to the extent that maximises joint profits from their diverse resources.

The complexities of such a wildlife trade regime arise with the incentives that it creates for appropriating the 'premium' without incurring the investment. Any state would have an incentive to attempt to sell diverse resources to the consumer-states that were the result of another state's investments in its natural habitats. This implies the contruction of an elaborate set of conditions for regulating trade between producer and the consumer-states, essentially individual supplier-states quotas based upon the natural habitats that they maintain and their capacity for production.

A wildlife trade regime should operate in this fashion because it is essential that there be price discrimination between those supplier-states investing and those not investing in their diverse resource bases. The formation of producer cooperatives consisting solely of the former creates a premium to that membership, and thus creates incentives for non-investing states to change strategies in this regard.

The current system under CITES does not adequately discriminate between investing and non-investing states in regard to legitimate wildlife trade, and its blanket 'bans' on endangered wildlife trade constitute a disproportionate punishment for the investing states (since they are then unable to acquire the returns on the investments they made). They bolster the general impression throughout the developing world that diverse resources cannot be developed economically, and thus affirm the perception that conversion is the sole pathway to development. For these reasons, wildlife trade reforms of the sort indicated above are essential for biodiversity conservation.

10.3.3   Informational Resource Rights

The most direct approach to the regulation of extinction would be the creation of an international mechanism for rewarding investments in the generation of the non-appropriable services of diverse resources. In a first-best world, this would take the form of a mechanism which renders the non-appropriable services from diverse resources appropriable.

It is generally believed that intellectual property regimes perform this latter function, when in fact they actually perform the former. That is, the function of an intellectual property regime is to create a mechanism for rewarding investments in the generation of non-appropriables, not the appropriation of these services. It accomplishes this by giving 'exclusive territories' within which to market the products that result, in part, from the investments in non-appropriables. So, investments in research on the better solution to a specific engineering problem (with the potential for a wide range of applications) are rewarded by giving the investor rights in a few selected products that incorporate this information.

Information can result from investments of various forms. Although it is widely seen to flow from investments in human capital, it is also one of the fundamental services to flow from investments in diverse resources. This information also serves as an input into the creation of many useful products. Therefore, it is perfectly consistent to speak of extending intellectual property regimes to encompass rewards for naturally-generated information, as well as human-generated.

Informational resource rights require the creation of an international agreement to award rights in product markets to those states successfully investing in the retention of diverse resource stocks for the purpose of information generation. Just as with patent competitions, states would have to compete for these awards by virtue of the demonstration of their retention of diverse habitats and their development of the capacities to effectively prospect within these habitats. In a periodic competition, the successful states would be awarded rights in specific chemical combinations that can be traced to prospecting.

In essence, informational resource rights constitute an alternative to the idea of 'transferring technology'. New institutions provide for the possibility that each part of the world may invest in its own comparative advantage, rather than having to integrate for enhanced appropriability. That is, at present, a developing country that wishes to capture the informational value of its diverse resources must become fully integrated vertically, because the exclusive rights do not attach before the final consumer product is developed. The idea of an informational resource right is to allow developing countries to specialise in the identification of active chemical combinations (shown to be derived from their stocks of diverse resources) without the necessity of proceeding to the development of the final consumer product. Then these informational resource rights (in the specific chemical combination) may be licensed out to various other users nearer to the product market, for investigation and potential development into marketable products. Since the values of the informational resource rights are wholly market-dependent, there is no value to the conferment of these rights unless the owner-state has created that value in the discovery of the chemical combination.

Informational resource rights are of particular importance because they go to the core of the biodiversity problem, i.e. the existence of inappropriable services. They are also important because they could provide a first example of an institution that would encourage new forms of development through the utilisation of diverse resources.

10.4 CONCLUSION

The international regulation of extinction is simply the regulation of the global conversion process. Just as decisions taken by individuals in a decentralised society are able to generate environmental problems, decisions taken by states in a multinational world can have the same effect. International institution-building is a necessity on account of these externalities.

As the same development process continues across the face of the earth, the urgency of international institution-building emerges. The costliness of conversions is increasing, and reaches potentially unbounded levels of costliness as the last remaining diversity is depleted.

This is because diversity is one of those systemic resources for which there are no human substitutes. It is valuable in part because it is the product of a specific process – four billion years of the evolutionary process. Humans may be able to create substitutes for the information and insurance services that evolution has generated, but it cannot recreate this specific process or its product.

For this reason, the developed world must act, and act in a very specific fashion. It must invest in international institutions that foster development down diverse pathways. Although the stream of funding for these institutions may have the appearance of a one-way street, the unidirectional character of this flow represents an attempt to create the other half of a two-way flow, i.e. the compensation required for providing the services of global diversity. This book has attempted to describe the nature of these services, and to develop the nature of the institutions that might channel the value of these services to its providers. The implementation of international agreements to this effect is what is required for the international regulation of extinction.

Bibliography

Acharya, R. (1991). 'Patenting of Biotechnology: GATT and the Erosion of the World's Biodiversity', *Journal of World Trade*, 25(6): 71–87.

Akerlof, G. (1970). 'The Market for Lemons', *Quarterly Journal of Economics*, 84(3): 488.

Alchian, A. and Demsetz, H. (1973). 'The Property Rights Paradigm', *Journal of Economic History*, 33: 16–27.

Anderson, J. and Hazell, P., (1989). *Variability in Grain Yields*, World Bank: Washington, DC.

Aoki, M. (1976). *Optimal Contol and System Theory in Dynamic Economic Analysis*. North Holland: New York.

Arrow, K. (1962). 'The Economic Implications of Learning by Doing', *Review of Economic Studies*, 29: 155–73.

Arrow, K. (1962). 'Economic Welfare and the Allocation of Resources for Invention', in R. Nelson (ed.), *The Rate and Direction of Inventive Activity*, National Bureau of Economic Research, Princeton University Press.

Arrow, K. (1964). 'The Role of Securities in the Optimal Allocation of Risk Bearing', *Review of Economic Studies*, 29: 155–73.

Arrow, K. (1971). *Essays in the Theory of Risk-Bearing*, North Holland: Amsterdam.

Arrow, K., and Fisher, A. (1974). 'Environmental Preservation, Uncertainty and Irreversibility', *Quarterly Journal of Economics*, 88: 312–19.

Arrow, K., and Kurz, M. (1970). *Public Investment, the Rate of Return and Optimal Fiscal Policy*, Johns Hopkins Press: Baltimore.

Atkinson. I. (1989). 'Introduced Animals and Extinctions', in Western, D. and Pearl, M. (eds), *Conservation for the Twenty-first Century*, Oxford University Press: Oxford.

Barbier, E., Burgess, J., Swanson, T. and Pearce, D. (1990). *Elephants, Economics and Ivory*, Earthscan: London.

Barret, S. 1989. 'On the Nature and Significance of International Environmental Agreements', mimeo, London Business School.

Barton, N. (1988). 'Speciation' in Myers, A. and Giller, P. *Analytical Biogeography*, Chapman and Hall: London.

Barzel, Y. (1991). *Economic Analysis of Property Rights*, Cambridge University Press: Cambridge.

Bell, R. and McShane-Caluzi, E., (eds) (1984). *Conversation and Wildlife Management in Africa* US Peace Corps: Washington, DC.

Berck, R. (1979). 'Open Access and Extinction', *Econometrica*, 47: 877–82.

Berkes, F. (ed.) (1989). *Common Property Resources Ecology and Community Based Sustainable Development*, Belhaven: London.

Binswanger, H. (1989). *Brazilian Policies that Encourage Deforestation in the Amazon*, Environment Department, World Bank, Working Paper No. 16, Washington DC.

265

Bjorndal, T., and Conrad, J. (1987). 'The Dynamics of an Open Access Fishery', *Canadian Journal of Economics*, 20(1): 74–85.

Boulding, K. (1981). *Ecodynamics*, Sage: London.

Boylan, E. (1979). 'On the Avoidance of Extinction in One-Sector Growth Models'. *Journal of Economic Theory*, 20(2): 276–9.

Brookshire, D. S., Eubanks, L. S. and Randall, A. (1983). 'Estimating Option Prices and Existence Values for Wildlife Resources', *Land Economics* 59: 1–15.

Browder, J. (1988). 'Public Policy and Deforestation in the Brazilian Amazon' in Repetto, R. and Gillis, M. (eds), *Public Policies and the Misuse of Forest Resources*, Cambridge University Press: Cambridge.

Brown, G. and Goldstein, J. (1984). 'A Model for Valuing Endangered Species', *Journal of Environmental Economics and Management*, 11: 303–9.

Brown, J. (1988), 'Species Diversity', in Myers, A. and Giller, P. *Analytical Biogeography*, Chapman and Hall: London.

Caughley, G. and Goddard, J. (1975). 'Abundance and distribution of elephants in Luangwa Valley, Zambia', *African Journal of Ecology*, 26: 323–7.

Cheung, S. (1970). 'The Structure of a Contract and the Theory of a Non-Exclusive Resource', *Journal of Law and Economics*, 13(1): 49–70.

Cheung, S. (1983). 'The Contractual Nature of the Firm', *Journal of Law and Economics*, 26(1): 1–21.

CIAT, (1981). Report on the Fourth IRTP Conference in Latin America, Cali, Colombia.

Ciriacy-Wantrup, S. and Bishop, R. (1975). 'Common Property as a Concept in Natural Resources Policy', *Natural Resources*, 15: 713–27.

Clark, C. (1973). 'Profit Maximisation and the Extinction of Animal Species', *Journal of Political Economy*, 81(4): 950–61.

Clark, C. (1973). 'The Economics of Overexploitation', *Science* 181: 630–4.

Clark, C. (1976). *Mathematical Bioeconomics: The Optimal Management of Renewable Resources*, John Wiley: New York.

Clark, C. and Munro, G. (1978). 'Renewable Resources and Extinction: Note', *Journal of Environmental Economics and Management*, 5(2): 23–9.

Clark, C., Clarke, F. and Munro, G. 1979. 'The Optimal Exploitation of Renewable Resource Stocks: Problems of Irreversible Investment', *Econometrica*, 47: 25–49.

Conrad, J., (1980). 'Quasi-Option Value and the Expected Value of Information', *Quarterly Journal of Economics*, 94: 813–20.

Conrad, J. and Clarke, C., (1987). *Natural Resource Economics*, Cambridge University Press: Cambridge.

Cox, C. and Moore, P. (1985). *Biogeography: An Ecological and Evolutionary Approach*, Blackwell: Oxford.

Cropper, M. (1988). 'A Note on the Extinction of Renewable Resources', *Journal of Environmental Economics and Management*, 15(1): 25–30.

Cropper, M., Lee, D., Pannu, S. (1980). 'The Optimal Extinction of a Renewable Natural Resource', *Journal of Environmental Economics and Management*, 6(4): 49–55.

Cumming, D., Du Toit, R. and Stuart, S., (1990). *African Elephants and Rhinos: Status Survey and Conservation Action Plan*, International Union for the Conservation of Nature: Gland.

Cyert, R., and DeGroot, M. (1987). 'Sequential Investment Decisions', in R. Cyert and M. DeGroot, *Bayesian Analysis and Uncertainty in Economic Theory*, Chapman and Hall: London.

Daly, H. (ed.) (1992). *Toward a Steady-State Economy* (2d edn), Island Press: Washington, DC.

Dandy, J. (1981). 'Magnolias', in B. Hora (ed.) *The Oxford Encylopedia of Trees of the World*, Oxford University Press: Oxford.

Darmstadter, J. (ed.)(1990). *Global Development and the Environment: Perspectives on Sustainability*, Resources for the Future: Washington, DC.

Dasgupta, P. (1969). 'On the Concept of Optimum Population', *Review of Economic Studies,* 36(3)(107): 295–318.

Dasgupta, P. (1982). *The Control of Resources*, Blackwell: Oxford.

Dasgupta, P. and Heal, G. (1974). 'The Optimal Depletion of Exhaustible Resources', *Review of Economic Studies*, Symposium Issue on Depletable Resources, 3–28.

Dasgupta, P. and Heal, G. (1979). *Economic Theory and Exhaustible Resources*, Cambridge University Press: Cambridge.

Dasgupta, P. and Stiglitz, J. (1988). 'Learning-By-Doing, Market Structure and Industrial and Trade Policies', *Oxford Economic Papers*, 40: 246–68.

David, P. (1985). 'Clio and the Economics of QWERTY', *American Economic Review*, Papers and Proceedings, 75: 332–7.

Davis, R., (1985). 'Research Accomplishments and Prospects in Wildlife Economics', *Transactions of N. American Wildlife and Natural Resource Conference*, 50: 392–8.

Davis, S. (1982). 'The Taming of the Few', *New Scientist*, 95: 697–700.

Dawkins, R. (1986). *The Blind Watchmaker*, Longman Scientific and Technical: Harlow.

Demsetz, H. (1967). 'Toward a Theory of Property Rights', *American Economic Review*, 57: 34–48.

Diamond, J. (1984). 'Normal extinctions of isolated populations', in Nitecki, M. (ed.). *Extinctions* University of Chicago: Chicago.

Diamond, J. (1989). 'Overview of Recent Extinctions', in Western, D. and Pearl, M. (eds). *Conservation for the Twenty-first Century*, Oxford University Press: Oxford.

Diamond, J. and Case, T. (eds) (1986). *Community Ecology*, Harper & Row: New York.

Dixit, A. (1992). 'Investment and Hysteresis', *The Journal of Economic Perspectives*, 6: 107–32.

Douglas-Hamilton, I. (1989). 'Overview of Status and Trends of the African Elephant', in Cobb, S. (ed.). *The Ivory Trade and the Future of the African Elephant,* Report of the Ivory Trade Review Group to the CITES Secretariat.

Duvick, D. N. (1984). 'Genetic Diversity in Major Farm Crops on the Farm and in Reserve', *Economic Botany* 38: 161–78.

Duvick, D. N. (1986). 'Plant Breeding: Past Achievements and Expectations for the Future', *Economic Botany* 40: 289–97.

Duvick, D. N. (1989). 'Variability in U.S. Maize Yields', in Anderson, J. and Hazell, P. (eds). *Variability in Grain Yields*, World Bank: Washington, DC.

Ehrlich, P. (1988). 'The Loss of Diversity: Causes and Consequences', in Wilson, E. O. (ed.), *Biodiversity*, op. cit.

Ehrlich, P. and Ehrlich, A. (1981). *Extinction*, Random House: New York.

Eltringham, S. K. (1984). *Wildlife Resources and Economic Development*, John Wiley: New York.

Farnsworth, N. (1988). 'Screening Plants for New Medicines', in Wilson, E. (ed.), *Biodiversity*, National Academy of Sciences: Washington, DC.

Farnsworth, N. and Soejarto, D. (1985). 'Potential Consequences of Plant Extinction in the United States on the Current and Future Availability of Prescription Drugs', *Economic Botany*, 39(3).

Findeisen, C. (1991). *Natural Products Research and the Potential Role of the Pharmaceutical Industry in Tropical Forest Conservation*. Rainforest Alliance: New York.

Fisher, A. and Hanneman, W. M. (1985). 'Endangered Species: The Economics of Irreversible Damage', in D. Hall, N. Myers, and N. Margaris (eds), *Economics of Ecosystem Management*, W. Junk Publishers: Dordrecht.

Fisher, A. and Krutilla, J. (1985). 'Economics of Nature Preservation', in A. Kneese and J. Sweeney (eds), *Handbook of Natural Resource and Energy Economics*, Elsevier: Amsterdam.

Fisher, A., Krutilla, J. and Cicchetti, C. (1972). 'Alternative Uses of Natural Environments: The Economics of Environmental Modification', in Krutilla, J. (ed.), *Natural Environments: Studies in Theoretical and Empirical Analysis*, Johns Hopkins University Press: Baltimore.

Fisher, A., Krutilla, J. and Cicchetti, C. (1972b). 'The Economics of Environmental Preservation: A Theoretical and Empirical Analysis', *American Economic Review*, 62: 605–19.

Fisher, A., Krutilla, J. and Cicchetti, C. (1974). 'The Economics of Environmental Preservation: Further Discussion', *American Economic Review*, 64: 1030–9.

Flint, M. E. S. (1990). 'Biodiversity: Economic Issues'. Unpublished paper for the Overseas Development Administration, London.

Foy, G. and Daly, H. (1989). *Allocation, Distribution and Scale as Determinants of Environmental Degradation: Case Studies of Haiti, El Salvador and Costa Rica*, World Bank, Environment Department Working Paper No. 19, Washington, DC.

Freeman, A. M. (1984). 'The Quasi-Option Value of Irreversible Development', *Journal of Environmental Economics and Management*, 11: 292–5.

Futuyma, D. (1986). *Evolutionary Biology*, Sinauer: Sunderland, MA.

Gadgil, M and Iyer, P. (1988). 'On the Diversification of Common Property Resource Use in Indian Society', in Berkes, F. (ed.), *Common Property Resources: Ecology and Community Based Sustainable Development*, Belhaven: London.

Gentry, A. (1982). 'Patterns of Neotropical Plant Species Diversity', in Prance, G. (ed.). *Biological Diversification in the Tropics*, Columbia University Press: New York.

Godfrey, M. (1985). 'Demand for Commodities', in Rose, T. (ed.), *Crisis and Recovery in Sub-Saharan Africa*, Paris: OECD.

Gordon, H. S. (1954). 'The economic theory of a common-property resource: the fishery', *Journal of Political Economy*, 62: 124–42.

Gould, J. R. (1972). 'Extinction of a Fishery by Commercial Exploitation: A Note', *Journal of Political Economy*, 80(5): 1031–8.

Grossman, S. and Hart, O. (1987). 'Vertical Integration and the Distribution of Property Rights', in Razin, A. and Sadka, E. (eds), *Economic Policy in Theory and Practice*, London: Macmillan.

Hanemann, M. (1989). 'Information and the Concept of Option Value', *Journal of Environmental Economics and Resource Management*, 16: 23–7.

Hanks, J. (1972). 'Reproduction of elephant in the Luangwa Valley, Zambia', *Journal of Reproduction and Fertility*, 30: 13–26.

Hardin, G. (1960). 'The Competitive Exclusion Principle', *Science*, 131: 1292–7.

Hardin, G. (1968). 'The tragedy of the common', *Science*, 162: 1243–8.

Hargrove, T., Cabanilla, V. and Coffman, W. (1988). 'Twenty Years of Rice Breeding', *Bioscience* 38: 675–81.

Harrington, W. and Fisher, A. (1982). 'Endangered Species, in Portney, P. (ed.), *Current Issues in Natural Resource Policy*, Resources for the Future: Washington, DC.

Hart, O. and Moore, J. (1990). 'Property Rights and the Nature of the Firm', *Journal of Political Economy*, 98(6): 1119–58.

Hartwick, J. (1977). 'Intergenerational Equity and the Investing of Rents from Exhaustible Resources', *American Economic Review*, 67(5): 972–84.

Hartwick, J. (1978). 'Investing Returns from Depleting Renewable Resource Stocks and Intergenerational Equity', *Economic Letters*, 1: 85–8.

Hays, J., Imbrie, J. and Shackleton, N. (1976). 'Variations in the earth's orbit: Pacemaker of the Ice Ages', *Science*, 213: 1095–6.

Hazell, P. (1984). 'Sources of Increased Instability in Indian and U.S. Cereal Production', *American Journal of Agricultural Economics*, 66.

Hazell, P. (1989). 'Changing Patterns of Variability in World Cereal Production', in Anderson, J. and Hazell, P. *Variability in Grain Yields*, World Bank: Washington DC.

Hazell, P. B. R., Jaramillo, M. and Williamson, A. (1990). 'The Relationship between World Price Instability and the Prices Farmers Receive in Developing Countries', *Journal of Agricultural Economics*, 41, No. 2

Heal, G. (1975). 'Economic Aspects of Natural Resource Depletion', in Pearce, D. and Rose, J. (eds), *The Economics of Depletion*, Macmillan: London.

Henry, C. (1974a). 'Investment Decisions Under Uncertainty: The Irreversibility Effect', *American Economic Review*, 64: 1006–12.

Henry, C. (1974b). 'Option Values in the Economics of Irreplaceable Assets', *Review of Economic Studies*, Symposium Issue on Depletable Resources, 89–104.

Hobbelink, H. (1991). *Biotechnology and the Future of World Agriculture*, Zed Brooks: London.

Hoel, M. 1990. 'Efficient International Agreements for Reducing Emissions of CO_2' mimeo, Department of Economics, University of Oslo.

Holdgate, M., Kassas, M., White, G. (eds). (1982). *The World Environment 1972–1982*, United Nations Environmental Programme: Nairobi.

Honneger, R. (1981). 'List of Amphibians and Reptiles either known or thought to have become extinct since 1600', *Biological Conservation*, 19: 141–58.

Hotelling, H. (1931). 'The economics of exhaustible resources', *Journal of Political Economy*, 39: 137–75.

Howard, N. (1991). *Legal Protection of Biotechnology within the European Community with Reference to Environmental Protection*, unpublished dissertation, University of London.

Iltis, H., (1988). 'Serendipity in the Exploration of Biodiversity', in Wilson, E. (ed.), *Biodiversity*, National Academy of Sciences: Washington, DC.

International Institute for Environment and Development and World Resources Institute (1989). *World Resources 1988–89*, Basic Books: New York.

International Monetary Fund. (1988). *World Economic Outlook 1988*, Washington, DC.

IUCN Environmental Law Centre. (1985). *African Wildlife Laws*, IUCN: Gland.

IUCN Environmental Law Centre. (1986). *Latin American Wildlife Laws*, IUCN: Gland.

Ivory Trade Review Group (ITRG). (1989). *The Ivory Trade and the Future of the African Elephant*, Report to the Conference of the Parties to CITES, Lausanne.

Johansson, P.-O. (1987). *The Economic Theory and Measurement of Environmental Benefits*, Cambridge University Press: Cambridge.

Juma, C. (1989). *The Gene Hunters*, Zed Books: London.

Kaldor, N. and Mirlees, J. (1962). 'A New Model of Economic Growth', *Review of Economic Studies*, 29: 174–92.

Kemp, M. and Long, N. (1984). *Essays in the Economics of Exhaustible Resources*, North Holland: Amsterdam.

Kiss, A. (ed.). (1990). *Living With Wildlife*, draft report of World Bank Environment Division, World Bank: Washington, DC.

Krautkramer, J. (1985). 'Optimal Growth, Resource Amenities and the Preservation of Natural Environments', *Review of Economic Studies*, 52: 153–70.

Krugman, P. (1979). 'A Model of Innovation, Technology Transfer and the World Distribution of Income', *Journal of Political Economy*, 79: 253–66.

Krutilla, J. V. (1967). 'Conservation Reconsidered', *American Economic Review* 57: 777–86.

Krutilla, J. V. and Fisher, A. (1975). *The Economics of Natural Environments*, Johns Hopkins University Press: Baltimore.

Kydland, F. and Prescott, E. (1977). 'Rules Rather than Discretion: The Inconsistency of Optimal Plans', *Journal of Political Economy*, 85(3): 471–91.

Leader-Williams, N., and Albon, S. (1988). 'Allocation of resources for conservation', *Nature*, 336, 533–5.

Ledec, G., Goodland, R., Kirchener, J. and Drake, J. (1985). 'Carrying Capacity, Population Growth and Sustainable Development', in D. Mahar (ed.), *Rapid Population Growth and Human Carrying Capacity*, World Bank Working Paper 690, World Bank: Washington, DC.

Lee, R. (1991). 'Comment: The Second Tragedy of the Commons', in K. Davis and M. Bernstam (eds), *Resources, Environment and Population: Present Knowledge and Future Options*, Princeton University Press: Princeton.

Leith, H. and Whittaker, R. (1975). *Primary Productivity of the Biosphere*, Springer-Verlag: New York.

Lewin, R. (1983). 'What Killed the Giant Land Mammals?', *Science*, 221: 1269–71.

Libecap, G. (1990). *Contracting for Property Rights*, Cambridge University Press: Cambridge.

Lovejoy, T. (1980). 'A Projection of Species Extinctions', in Barney, G. (ed.), *The Global 2000 Report to the President*, Council on Environmental Quality: Washington, DC.

Lucas, R. (1988). 'On the Mechanics of Economic Development', *Journal of Monetary Economics*, 22: 3–22.

Lugo, A. (1986). 'Estimating Reductions in the Diversity of Tropical Forest Species', in E. O. Wilson (ed.),*Biodiversity*, National Academy of Sciences: Washington, DC.

Luxmoore, R., Caldwell, J. and Hithersay, L., (1989). 'The Volume of Raw Ivory Entering International Trade from African Producing Countries from 1979 to

1988', in Cobb, S. (ed.), *The Ivory Trade and the Future of the African Elephant*, Report of the Ivory Trade Review Group to the CITES Secretariat.

Lynch, J. (1988). 'Refugia', in Myers, A. and Giller, P. (eds) *Analytical Biogeography*, Chapman and Hall: London.

Lyster, S. (1985). *International Wildlife Law*, Grotius: London.

Mabberley, D. (1992). 'Coexistence and Coevolution', in *Tropical Forest Ecology*, Chapman & Hall: New York.

MacArthur, R. and Wilson, E. (1967). *The Theory of Island Biogeography*, Princeton University Press: Princeton.

McKelvey, R. (1987), 'Fur Seal and Blue Whale: The Bioeconomics of Extinction', in Cohen, Y. (ed.), *Applications of Control Theory in Ecology*, Lecture Notes in Biomathematics Series, no. 73: New York.

McNeely, J., Miller, K., Reid, W., Mittermeier, R. and Werner, T. (1990). *Conserving the World's Biological Diversity*, IUCN: Gland, Switzerland.

McNeely, Jeffrey A. (1988). *Economics and Biological Diversity: Developing and Using Economic Incentives to Conserve Biological Resources*. IUCN: Gland.

Mahar, D. (1989). *Government Policies and Deforestation in Brazil's Amazon Region*, World Bank: Washington, DC.

Marks, S. (1985). *The Imperial Lion: Human Dimensions of Wildlife Management in Africa*, Colorado: Westview Press.

Marshall, L. (1988). 'Extinction', in Myers, A. and Giller, P. (eds) *Analytical Biogeography*, Chapman and Hall: London.

Mathewson, F. and Winter, R. (1986). 'The Economics of Vertical Restrains in Distribution', in Mathewson, F. and Stiglitz, J. (eds), *New Developments in the Analysis of Market Structure*, MIT Press: Boston.

Meadows, D. H., Meadows, D. L., Jorgen, Randers and Behrens, W., (1974). *The Limits to Growth*, Universe Books: New York.

Miller, J. and Lad, F. (1984). 'Flexibility, Learning and Irreversibility in Environmental Decisions', *Journal of Environmental Economics and Management*, 11: 161–72.

Mittermeier, R. (1988). 'Primate Diversity and the Tropical Forest', in Wilson E. O., (ed.), *Biodiversity*, National Academy of Sciences: Washington, DC.

Muscat, R. (1985). 'Carrying Capacity and Rapid Population Growth: Definition. Cases and Consequences', in Mahar, D. (ed.), *Rapid Population Growth and Human Carrying Capacity*, World Bank Working Paper 690, World Bank: Washington, DC.

Myers, A. and Giller, P. (1988). *Analytical Biogeography*, Chapman and Hall: London.

Myers, N. (1979). *The Sinking Ark. A Look at the Problem of Disappearing Species*. Pergamon: New York.

Myers, N. (1983). *A Wealth of Wild Species*, Westview: Boulder.

Myers, N. (1984). *The Primary Source*, Norton: New York.

NAS. (1972). *Genetic Vulnerability of Major Form Crops*, National Research Council, NAS: Washington, DC.

Norgaard, R. (1986). 'The Rise of the Global Exchange Economy and the Loss of Biological Diversity', in Wilson, E. (ed.), *Biodiversity*, National Academy of Sciences: Washington, DC.

Norton, B. (ed.) (1986). *The Preservation of Species*, Princeton University Press: Princeton, New Jersey.

Office of Technology Assessment (1988). *Technologies to Maintain Biological Diversity*, Lippincott: Philadelphia.

Oldfield, M. (1984). *The Value of Conserving Genetic Resources*, US Department of Interior: Washington, DC.

Olson, S. (1989). 'Extinction on Islands: Man as a Catstrophe', in Western, D. and Pearl, M. (eds), *Conservation for the Twenty-first Century*, Oxford University Press: Oxford.

Ostrom, E. (1990). *Governing the Commons*, Cambridge University Press: Cambridge.

Panayotou, T. (1989). *The Economics of Environmental Degradation Problems, Causes and Responses*, Harvard Institute for International Development, Cambridge, Massachusetts (reprinted in Markandya, A. and Richardson, J. (eds) (1992), *Environmental Economics*, Earthscan: London.

Pearce, D. W. (1988). 'The Sustainable Use of Natural Resources in Developing Countries', in Turner, R. K. (ed.), *Sustainable Environmental Management: Principles and Practice*, Belhaven Press, London.

Pearce, D. (1989). *The Ivory Trade 1950–1978*, in IRTG (1989).

Pearce, D. W. (1991). 'An Economic Approach to Saving the Tropical Forests' in Helm, D. (ed.), *Economic Policy Towards the Environment*, Blackwell: Oxford.

Pearce, D. W., Barbier, E. and Markandya, A. (1990). *Sustainable Development Economics and Environment in the Third World*, Edward Elgar, London.

Perrings, C. (1989). 'An Optimal Path to Extinction? Poverty and Resource Degradation in the Open Economy', *Journal of Development Economics*, 30(1): 1–24.

Perrings, C. (1991). 'Ecological Sustainability and Environmental Control', *Structural Change and Economic Dynamics*, 2: 275–95.

Perrings, C. (1992). 'Biotic Diversity, Sustainable Development and Natural Capital', Paper presented to the International Society for Ecological Economics, Stockholm, Sweden, August, 25 pp.

Peters, C., Gentry, A. and Mendelsohn, R. (1990). 'Valuation of an Amazonian Rainforest', *Nature*, 339: 655–7.

Peterson, W. and Randall, A. (1984). *Valuation of Wildland Resource Benefits*, Westview Press: Boulder, Colorado.

Phelps, E. (1961). 'The Golden Rule of Accumulation: A Fable of Growth-men', *American Economic Review*, 57: 89–99.

Plucknett, D. L. Smith, N. J. H. (1986). 'Sustaining Agricultural Yields', *BioScience* 36: 40–5.

Plucknett, D. L., Smith, N. J. H., Williams, J. T. and Murthi Anishetty, N. (1983). 'Crop Germplasm Conservation and Developing Countries', *Science* 8: 163–9.

Plucknett, D. L., Smith, N. J. H., Williams, J. T. and Murthi Anishetty, N. (1987). *Gene Banks and the World's Food*. Princeton: Princeton University Press.

Pharmaceutical Manufacturers Association: Annual Survey Report 1988–1990, cited in World Conservation Monitoring Centre (1992). *Global Biodiversity 1992*, WCMC: Cambridge.

Pindyck, R. (1978a). 'The Optimal Exploration and Production of Nonrenewable Resources', *Journal of Political Economy*, 86: 841–61.

Pindyck, R. (1978b). 'Gains to Producers from the Cartelization of Exhaustible Resources', *Review of Economics and Statistics*, 60: 238–51.

Pindyck, R. (1991),'Irreversibility, Uncertainty and Investment', *Journal of Economic Literature*, 29: 1110–48.

Pinstrup-Anderson, P. and Hazell, P. R. B. (1985). 'The Impact of the Green Revolution and the Prospect for the Future', *Food Reviews International*, 1(1): 1–12.

Plotkin, M. (1988). 'The Outlook For New Agricultural and Industrial Products from the Tropics', in E. O. Wilson, 1988.

Porter, R. C. (1982). 'The New Approach to Wilderness Preservation through Cost-Benefit Analysis', *Journal of Environmental Economics and Management*, 9: 59–80.

Prescott-Allan, R. and Precott-Allan, C. (1983). *Genes from the Wild*, Earthscan: London.

Principe, P. (1991). 'Valuing the Biodiversity of Medicinal Plants', in Akerle, O., Heywood, V. and Synge, V., *The Conservation of Medicinal Plants*, Cambridge University Press: Cambridge.

Quian, Y. (1992). 'Equity, Efficiency and Incentives in a Large Economy', *Journal of Comparative Economics*, 27–46.

Ramamohan Rao, T. V. S. (1991). 'Efficiency and Equity in Dynamic Principal-Agent Problems', *Journal of Economics*, 55(1): 17–41.

Randall, A. (1991). 'Total and Nonuse Values', in Braden, B. and Kolstad, D. (eds), *Measuring the Demand for Environmental Quality*, Amsterdam: Elsevier Science Publishers BV.

Raup, D. (1988). 'Diversity Crises in the Geological Past', in Wilson, E. O. (ed.), *Biodiversity*, National Academy Press: Washington, DC.

Ready, R. and Bishop, R. (1991). 'Endangered Species and the Safe Minimum Standard', *American Journal of Agricultural Economics*, 73(2): 309–12.

Reid, W. and Miller, K., (1989). *Keeping Options Alive*, World Resources Institute: Washington, DC.

Renewable Resources Assessment Group. (1989). 'The Impact of the Ivory Trade on the African Elephant Population', in Cobb, S. (ed.), *The Ivory Trade and the Future of the African Elephant*, Report of the Ivory Trade Review Group to the CITES Secretariat.

Repetto, R. (ed.) (1985). *The Global Possible*, Yale University Press: New Haven.

Repetto, R. (1986a). *World Enough and Time*, Yale University Press: New Haven.

Repetto, R., (1986b). 'Soil Loss and Population Pressure on Java', *Ambio*, 15: 14–20.

Repetto, R. (1988). *Economic Policy Reform for Natural Resource Conservation*, World Bank, Environment Department Working Paper No. 4, Washington, DC.

Repetto, R. and Gillis, M. (1988). *Public Policies and the Misuse of Forest Resources*, Cambridge University, Press: Cambridge.

Rhoades, R. (1991). 'The World's Food Supply at Risk', *National Geographic* 179(4): 74–103.

Robinson, W. and Bolen, E., (1989). *Wildlife Ecology and Management*, (2d edn), Macmillan: New York.

Romer, P. (1986). 'Increasing Returns and Long Run Growth', *Journal of Political Economy*, 94: 1002–27.

Romer, P. (1987). 'Growth Based on Increasing Returns due to Specialisation', *American Economic Review*, Papers and Proceedings, 77: 56–62.

Romer, P. (1990a). 'Endogenous Technological Change', *Journal of Political Economy*, 98: 245–73.

Romer, P. (1990b). 'Are Nonconvexities Important for Understanding Growth?, *American Economic Review*, Papers and Proceedings, 80(2): 97–103.

Ruitenbeek, J. (1990). *Evaluating Economic Policies for Promoting Rainforest Conservation in Developing Countries*, unpublished Ph. D. dissertation, London School of Economics, University of London.

Samuelson, P. (1954). 'The Pure Theory of Public Expenditure', *Review of Economics and Statistics*, 36: 387–9.

Samuelson, P. (1976). 'Economics of Forestry in an Evolving Society', *Economic Inquiry*, 14: 466–92.

Sappington, D. (1991). 'Incentives in Principal–Agent Relationships', *Journal of Economic Perspectives*, 5(2): 45–66.

Schneider, R., *et al.* (1992). *Brazil: An Analysis of Environmental Problems in the Amazon*, World Bank: Washington, DC.

Scott, A. (1955). 'The Fishery: The Objectives of Sole Ownership', *Journal of Political Economy*, 63: 116–24.

Simberloff, D. (1986). 'Are we on the verge of an mass extinction in tropical rain forests?', in Elliot, D. (ed.), *Dynamics of Extinction*, John Wiley: New York.

Simon, J. (1992). *Population and Development in Poor Countries*, Princeton University Press: Princeton.

Sinn, H. (1982). 'The Economic Theory of Species Extinction: Comment on Smith', *Journal of Environmental Economics and Management*, 9(2): 82–90.

Smith, V. K. (1972). 'The Effects of Technological Change on Different Uses of Environmental Resources', in Krutilla, J. V. (ed.), *Natural Environments: Studies in Theoretical and Applied Analysis*, Johns Hopkins University Press: Baltimore.

Smith, V. K. (ed.) (1979). *Scarcity and Growth Reconsidered*, Johns Hopkins University Press: Baltimore.

Smith, V. K. (1987). 'Uncertainty, Benefit–Cost Analysis, and the Treatment of Option Value', *Journal of Environmental Economics and Management*, 14: 283–92.

Smith, V. K. and Krutilla, J. V. (eds) (1982). *Explorations in Natural Resource Economics*, Johns Hopkins University Press: Baltimore.

Smith, V. L. (1975). 'The Primitive Hunter Culture, Pleistocene Extinction and the Rise of Agriculture', *Journal of Political Economy*, 83: 727–55.

Smith, V. L. (1977). 'Control Theory Applied to Natural and Environmental Resources: An Exposition', *Journal of Environmental Economics and Management*, 4: 1–24.

Solbrig, O. (1991). 'The Origin and Function of Biodiversity', *Environment*, 33: 10–19.

Solow, R. (1963). *Capital Theory and the Rate of Return*, North Holland: Amsterdam.

Solow, R. (1974a). 'The Economics of Resources or the Resources of Economics', *American Economic Review*, 64: 1–12.

Solow, R. (1974b). 'Intergenerational Equity and Exhaustible Resources', *Review of Economic Studies*, Symposium Issue on Depletable Resources, 37–48.

Southgate, D. and Pearce, D. W. (1988). *Agricultural Colonisation and Environmental Degradation in Frontier Developing Economies*, World Bank, Environment Department Working Paper No. 9, Washington, DC.

Spence, M. (1975). 'Blue Whales and Applied Control Theory', in Gottinger, H. (ed.), *System Approaches and Environmental Problems*, Vandenhoeck: Gottingen.

Stanley, S. (1986). *Earth and Life through Time*, W. H. Freeman: London.

Stern, N. (1989). 'The Economics of Development: A Survey', *Economic Journal*, 99: 597–685.

Swallow, S. (1990). 'Depletion of the Environmental Bias for Renewable Resources', *Journal of Environmental Economics and Management*, 19(3): 281–96.

Swallow, S., Pooks, P. and Wear, D. (1990). 'Policy-Relevant Nonconvexities in the Production of Multiple Forest Benefits', *Journal of Environmental Economics and Management*, 19(3): 276–80.

Swaney, J. and Olson, P. (1992). 'The Economics of Biodiversity: Lives and Lifestyles', *Journal of Economic Issues*, 26(1): 1–25.

Swanson, T. (1989a). 'Policy Options for the Regulation of the Ivory Trade', in ITRG, *The Ivory Trade and the Future of the African Elephant*, Lausanne.

Swanson, T. (1989b). 'A Proposal for the Reform of the African Elephant Ivory Trade', London Environmental Economics Centre DP 89-04, International Institute for Environment and Development: London.

Swanson, T. (1990a). 'Conserving Biological Diversity', in Pearce, D. (ed.), *Blueprint 2: Greening the World Economy*, Earthscan: London.

Swanson, T. (1990b). 'Wildlife Utilisation as an Instrument for Natural Habitat Conservation: A Survey', London Environmental Economics Centre Discussion Paper 91-03.

Swanson, T. (1991a). 'The Environmental Economics of Wildlife Utilisation', in *Proceedings of the IUCN Workshop on Wildlife Utilisation*, IUCN: Gland.

Swanson, T. (1991b). 'Animal Welfare and Economics: The Case of the Live Bird Trade', in Edwards, S. and Thomsen, J. (eds), *Conservation and Management of Wild Birds in Trade*, Report to the Conference of the Parties to CITES, Kyoto, Japan.

Swanson, T. (1992a). 'Policies for the Conservation of Biological Diversity', in Swanson, T. and Barbier, E. (eds) (1992), *Economics for the Wilds: Wildlands, Wildlife*, Diversity and Development, Earthscan: London.

Swanson, T. (1992b). 'The Economics of a Biodiversity Convention', *Ambio*, 21: 250–7.

Swanson, T. (1992c). 'The Evolving Trade Mechanisms in CITES', *Review of European Community and International Environmental Law*, 1: 57–63.

Swanson, T. and Barbier, E. (eds) (1992). *Economics for the Wilds: Wildlands, Wildlife, Diversity and Development*, Earthscan: London.

Swanson, T. and Pearce, D. (1989). 'The International Regulation of the Ivory Trade – The Ivory Exchange', Paper prepared for the International Union for the Conservation of Nature: Gland.

Terborgh, J. (1974). 'Preservation of Natural Diversity: The Problem of Extinction Prone Species', *Bioscience* 24: 715–22.

Vitousek, P., Ehrlich, P., Ehrlich, A., and Matson, P. (1986). 'Human Appropriation of the Products of Photosynthesis', *Bioscience*, 36(6): 368–73.

Watson, D. (1988). 'The Evolution of Appropriate Resource Management Systems', in Berkes, F. (ed.), *Common Property Resources: Ecology and Community Based Management*, op. cit.

Weitzman, M. (1970). 'Optimal Growth with Scale Economies in the Creation of Overhead Capital', *Review of Economics Studies*, 37: 555–70.

Weitzman, M. (1976). 'Free Access versus Private Ownership as Alternative Systems for Managing Common Property', *Journal of Economic Theory*, 8: 225–34.

West, R. (1977). *Pleistocene Geology and Biology*, Longman: London.

Western, D. (1989). 'Population, Resources and Environment in the Twenty-first Century', in Western, D. and Pearl, M. (eds), *Conservation in the Twenty-first Century*, Oxford University Press: Oxford.

Western, D., and Pearl, M. (1989). *Conservation for the Twenty-first Century*, Oxford University Press: Oxford.

Wijnstekers, W. (1988). *The Evolution of CITES*, Secretariat of the Convention on International Trade in Endangered Species: Lausanne.

Williamson, O. (1975). *Markets and Hierarchies*, Free Press: New York.

Williamson, O. (1980). 'Transaction Cost Economics: The Governance of Contractual Relations', *Journal of Law and Economics*, 70: 356–72.

Williamson, O. (1983). 'Credible Commitments: Using Hostages to Support Exchange', *American Economic Review*, 73(4): 519.

Williamson, O. (1986). *The Economics of Capitalist Institutions*, Blackwell: Oxford.

Williamson, M. (1988). 'Relationship of Species Number to Area, Distance and other Variables', in Myers, A. and Giller, P. *Analytical Biogeography*, Chapman and Hall: London.

Wilson, E. (1988). *Biodiversity*, National Academy Press: Washington, DC.

Wilson, E. O. (1988). 'The Current State of Biological Diversity', in Wilson, E. (ed.), op. cit.

Witt, S. (1985). *Biotechnology and Genetic Diversity*, California Agricultural Lands Project: San Francisco.

World Bank (1989). *World Development Report*, Oxford University Press, Oxford.

World Bank (1990). *World Development Report*, World Bank: Washington, DC.

World Bank (1992). *World Development Report*, World Bank: Washington, DC.

World Conservation Monitoring Centre (WCMC), (1992). *Global Biodiversity*, Chapman and Hall: London.

World Resources Institute (1990). *World Resources 1990–1991*, Oxford University Press: Oxford.

Wright, H. (1970). 'Environmental Changes and the Origin of Agriculture in the Near East'. *Bioscience*, 20: 19–23.

Yang, X. and Wills, I. (1989). 'A Model Formalising the Theory of Property Rights', *Journal of Comparative Economics*, 14: 177–98.

Index